Math Mammoth
Discovery Word Problems

By Maria Miller

Contents

Contents

Introduction

Math Mammoth Discovery Word Problems is intended for students in 4th grade and above that need more practice with word problems. I especially aim to help students in grades 4-7 who have trouble with word problems and who may need to "go almost back to the beginning," so to speak, to learn to solve them.

The skills covered include addition, subtraction, multiplication, division, large numbers, divisibility, units of measurement, bar models in problem solving, average, geometry, fractions, and decimals. The problems require knowledge of fourth grade math, such as long division, long multiplication, knowledge of basic measurement units, and the usage of several operations. However, there are only a few worksheets dedicated to fractions and decimals, and the problems in general do not require calculations with fractions or decimals, with the exception of money amounts.

The book presents a variety of situations — many of them about animals and farming — so that students will have some fun as they look for and discover the solutions. Mathy the Mammoth, with his obsession with blueberries, is featured several times, and students will also go hiking and traveling around the world. There are also some challenging word problems (marked as such) and interesting puzzle corners.

Some worksheets have examples and some instruction related to problem solving. Students are often encouraged to write down the calculations they do. I suggest that you also encourage your student(s) to make a drawing or a sketch of the situation in the problem (if applicable), to help them understand the situation and to see what math operation is involved.

Overall, the problems in this collection require multiple steps, with the exception of question #1 in worksheet "Division Warm-Up", as it focuses on differentiating what math operation should be used.

If the problems in this book seem slightly too difficult, check out *Math Mammoth Foundational Word Problems*. The problems in it only require concepts and skills from 3rd grade math.

I hope that your students will discover that they indeed can solve word problems!

Maria Miller

Identify the Operation

When solving word problems, always read the problem carefully, and imagine in your mind what is happening. You can also draw or act out the situation to help you!

Ask yourself what the problem is asking for. Then, think about what math operation(s) match the situation. Here are some tips:

- **If you are finding the total amount of something**: multiplication or addition

- **If you know the total, or what the total used to be**: subtraction or division

- **If there are no equal-sized groups**: addition or subtraction

- **If there are equal-sized groups**: multiplication or division

These tips won't answer every scenario! So don't blindly choose an operation. Also, many word problems require several operations.

Example. Brian divided three pizzas into ninths. How many slices did he end up with?

The problem includes the word "divided". Is it about division? Imagine 3 pizzas in your mind, cut into ninths. This is a question about three equal-sized "groups" (pizzas), with 9 in each "group". To find the total number of slices, we *multiply*, not divide: $3 \times 9 = 27$. Brian got 27 slices.

1. Circle the equation(s) that match(es) the problem. Solve.

a. Randy buys a bicycle for $123. Now he has $94 left. How much did he have before?	$123 + 94 = \underline{\quad}$	$94 + \underline{\quad} = 123$
	$\underline{\quad} - 123 = 94$	$123 - \underline{\quad} = 94$
b. Priscilla pays $48 for eight mugs. How much was each mug?	$48 - 8 - 8 = M$	$48 + 8 = M$
	$6 \times M = 48$	$48 \div 8 = M$
c. Scarlet buys a $23 shirt and six snack bars for a total of $41. How much did each snack bar cost?	$41 - 23 - 6 = x$	$23 + 6 \times 41 = x$
	$6 \times S + 23 = 41$	$(41 - 23) \div 6 = S$

2. Solve. On the empty line, write a calculation or several for the problem.

a. Amy added three numbers and got a sum of 730. Two of the numbers were 260 and 128. What was the third number?

_____

The third number was _____.

b. David drove from Dallas to Houston, a distance of 239 miles, and back, twice in a month. How far did he drive?

_____

He drove _____ miles.

c. Elaine is making a quadruple batch of apple crisp for a party. Her recipe calls for nine apples. How many does she need?

_____

Elaine needs _____ apples.

d. Fiona is 131 cm tall and her little sister is 119 cm tall. What is the difference in their heights?

_____

Their heights are _____ cm apart.

e. Denise went shopping with $500 and came home with $249. How much did she spend?

_____

Denise spent _____.

f. There are a bunch of bikes and seven trikes in front of a daycare place. In total, those have 41 wheels. How many bikes are there?

_____

There are _____ bikes.

Addition and Subtraction

1. Circle the equation that fits the problem (x stands for the unknown number). Then solve for x.

a. Hunter had 197 drawings. Then he drew some more (x), and had 204.

$$197 + x = 204 \qquad 197 - x = 204$$

$x =$ _____

b. Mike caught a bunch of small fish in a lake, but a fox ate 15. Now he has 14 left.

$$15 - x = 14 \qquad x - 15 = 14$$

$x =$ _____

c. You buy two jumping ropes for $8.50 each and get $3 in change. What did you pay with?

$$x + 8.50 + 8.50 = 3 \qquad x - 8.50 - 8.50 = 3 \qquad 3 = 8.50 + 8.50 - x$$

$x =$ _____

> Sometimes, not all the information in a word problem is necessary. Don't let the extra numbers confuse you! Only use the numbers you need to solve the problem.

2. Solve.

a. Brent rides his bike for 60 minutes, going 22 km in an hour, and the next two days he rides his bike for 25 km and 18 km. What was the total distance he rode over those three days?

b. Jude earned $412 this summer. Stan earned $196 less than that, Perry earned $83 more than Stan, and John earned $13 more than Perry. How much did Perry earn?

c. Sam goes to the gym for half an hour on three days of the week, rides his bike for an hour on Thursday, jogs for 45 minutes on Saturday, and watches television for 50 minutes on the rest of the days. How many minutes does he spend exercising in a week?

How many hours and minutes is that?

3. Write an equation(s) for each problem, and then solve.

a. Maya planted one pine tree in her yard every year for four years. Now the tallest pine tree is 116 inches tall, and each pine tree after that is 22 inches shorter than the previous. How tall is the shortest pine tree?

b. A flock of 237 geese migrates to South Carolina one winter. The next year, the same flock has doubled in size, and the year after that, they have doubled again. How big is the flock now?

4. Match the situation to its equation!

a. Kayla has 123 paintings left after selling 48.

i. $123 - 48 = x$

b. Kayla sold all but 48 of her 123 paintings.

ii. $x - 48 = 123$

c. Kayla sold 48 of her 123 paintings.

iii. $123 - x = 48$

5. Write your own word problem! Choose one of the given topics.

Shane buys two shovels...   Cindy rides her horse for...   Mathy is eating blueberries...

What Is It Asking?

Many word problems take more than one step. Check first what the word problem is actually asking for. Don't blindly use the numbers in the problem, but think what type of information you would need in order to solve what the problem is asking for. You may need to first find out something else, before you can solve what the problem asks for.

Example. A SuperStar soccer ball costs $9.50. Another brand costs $12 more. Joe buys two of the SuperStar ones and one of the more expensive ones, and pays with $45. What is his change?

The problem is asking for Joe's change, but to find that out, we first need to find out the answer to the question, "What was the total cost of the balls?" And to find the cost of the balls, we first need to find the price of the more expensive ball.

So here, **Step 1** is to add $9.50 + $12 = $21.50, to get the price of the second ball.

Then **Step 2** is to find the total cost of the balls by adding $9.50 + $9.50 + $21.50 = $40.50.

Finally, **Step 3** is to find his change, which we get by subtracting: $45 − $40.50 = $4.50.

Joe got $4.50 in change.

1. Find a different question to ask first. Then solve.

a. Of 98 students, 43 are girls. How many more boys are there than girls?

First find out: _____

b. Cash has been saving up to buy a laptop for $989 and a mouse for $9.
He has $821. How much more does he need to earn before he can buy them?

First find out: _____

2. Solve. You can write more questions to solve first.

a. A store sells a blueberry-flavored energy bar for $3.75. Callie buys three of them and pays with $20. She gets $8.75 in change. Was that correct?

First find out: _____

b. One day, Charlotte the goose starts to build a nest, and she spends 6 hours doing that. Over the next six days, she continues to work on it, but for less time. Each of those days, she only works on the nest for a third of the time she spent the first day. How long does she take building her nest?

First find out: _____

c. The Fletcher family drove to a ski track to ski, and are now driving back. On their way home, they stopped at a gas station that is 29 miles from the ski track. By that point, they had driven a total of 84 miles (starting from home). How many more miles do they have to drive before they get home?

First find out: _____

3. These word problems have problems! Read them carefully. Find the problem, in each case.

a. Flora had 23 chickens. Then, two of them had chicks. One had 14 chicks. How many chickens and chicks does Flora have now?	**b.** Fiona is 142 cm and her little sister is 15 cm shorter. What is the difference in their heights?

Using Bar Models

Bar models can help you to see how the numbers in the problem relate to each other.

Example. In a herd of goats, there are 31 female kids and 28 male kids, and they make up half of the flock. The rest are adults. How many goats are there?

The bar represents the entire flock. We show the 31 and 28 as being half of the bar.

There are 31 + 28 = 59 kids, and this is half of the total. This means there are 59 + 59 = <u>118 goats</u> in total.

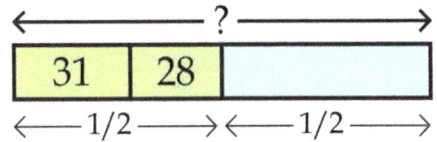

1. Solve. Draw a bar model for each problem.

a. The Trotters family runs a horse breeding farm. They have a total of 98 horses. The colors of the horses are: 21 bay, 13 black, 17 gray, and also some chestnut horses. How many horses are chestnut?	**b.** A store is expecting a shipment of 5,000 books. Three boxes of 650 have come. How many have yet to arrive?
c. Mathy the Mammoth is eating his breakfast. After feasting on 5 kg of huckleberries and 14 kg of blueberries, he is half-way through his breakfast. How much is he eating for breakfast?	**d.** Derek has 48 chickens. Half of them are red, 12 are black, eight are white, and the rest are gold. How many are gold?

2. Fill in the numbers from the problem in the bar model. Also, write an *x* in the model to signify what the problem is asking for. Then solve.

> The Richford family is on a 572-mile road trip. At a gas station 93 miles ago, Terry had calculated they had 321 miles left to go. How far have they traveled by now?

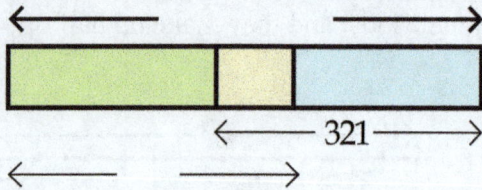

3. Draw a bar model to go along with each problem, and solve.

> **a.** Dallas had $584. He bought a puppy named Jet, and after that, he had $465 left.
> Then he put aside $120 of that to cover for any dog supplies and food in the near future.
> Considering this, how much money is he actually spending on Jet?

> **b.** Wayne's mom bought a subscription for Wayne to a math facts practice program.
> For the first two months, she only paid half of the normal price. Then she continued paying the full price for the next three months. For these five months, she paid a total of $44. What is the normal price of the program for one month?

> **c.** Nora is making necklaces for her friends. She uses 153 beads on one necklace and 96 on a second, and then uses all but the last 51 of her beads on the third necklace. She started with 400. How many did she use on the third necklace?

Money and Discounts

1. Solve.

a. Griffin bought a toy dump truck for $19 and a watch for $25. Now he has $41 left. How much did he have originally?

b. Helen had $80. Then she bought two sets of acrylic paints (both cost the same), and had $28 left. How much did one paint set cost?

c. A backpack that used to cost $63.55 is discounted by $7.90. How much is it now?

d. Rose buys three umbrellas for $11.69 each and pays with two $20-dollar bills. How much is her change?

e. Dad goes shopping with $531, buys some flyswatters for $4 each, spends $172 on groceries, and comes home with $327. How many flyswatters did he buy?

f. Brad decides to buy three flashlights for $14 each, but the seller gives him a discount. He ends up paying $36. How much was the discount *per flashlight*?

2. Solve.

a. Nadia and Marie are shopping for clothes. Marie wants a set of six shirts that costs $48 and Nadia wants a set of four shirts that costs $36. Which set has cheaper shirts?

b. If Liz is buying an expensive couch that costs $3,269.95 and paying in cash using $100 bills, how many bills does she need?

What would her change be?

c. At the annual sale of Willy's Wacky Warehouse, you get a $50 discount for every $200 worth of stuff you buy. Corey is buying a wheelbarrow for $99, a drill for $53.68, and some shovels for $9.50 each.

How many shovels should he get so he could get the discount?

What will be his final total with the discount?

Puzzle Corner   Potatoes cost $1.50 per pound. Also, you get a $2 discount if you buy at least $10 worth of potatoes. How many (whole) pounds of potatoes can you get with $11?

16

Going Hiking

This map shows a net of hiking trails. Refer to this map for all the problems in this lesson, and fill in the missing distances as necessary.

1. Lee only wants to see Rock That Looks Like A Bear, so he decides to take the shortest loop (ABEF). But at point E he sees an actual bear, and runs back to point B and then A. What distance in total did he cover?

   First, estimate it: _____

2. You're at Tim's Small Cave and want to take the shortest possible route to an exit (A, F, D, or J). Which way do you go? How long is the way you'll take?

3. The Lowe family starts at F and end in D. They cross five rivers, see three points of interest (indicated by stars on the map), and walk a distance of approximately 13,000 feet. What exact path did they take?

4. Which is longer: Shadow Trail, or the opposite path, JIHGEF? How much longer?

5. Ellie rides her horse Jasper down Shadow Trail. Then, she rides back the other way, taking the path JIHGEF, making a loop. If Jasper is going at an average of 6 miles per hour, *about* how many times can she ride around this loop within two hours? A mile is 5,280 feet. (Estimate using rounded numbers.)

First find out: _____

6. Choose your own hiking trail! Make sure it starts and ends at any of the exit points and passes at least two points of interest. Find out how long it is.

Thousands Upon Thousands

1. Fill with rounded numbers.

a. Mount McKinley is 20,310 feet tall, and Mount Everest is 29,035 feet tall.

Mount Everest is approximately _____ ft taller than Mount McKinley.

b. Rebecca earned $58,313 this year, which is about $_____.

She earned about $_____ each month.

c. Hans earned $3,198 this month, which is about $_____.

He would earn about $_____ in a year.

2. This chart shows how many berries Mathy the Mammoth has eaten so far this year.

	Jan	Feb	Mar	Apr	May	June	July
Blueberries	27,128	29,910	31,016	33,256	35,492	37,111	39,693
Blackberries	9,231	12,545	11,821	9,985	12,026	11,621	12,470
Strawberries	0	0	125	902	1,265	2,912	4,721

a. How many berries in total did Mathy eat in April?

b. In which month did Mathy eat about a fourth as many strawberries as blackberries? (Use rounded numbers.)

c. About how many strawberries did he eat in these seven months? Use numbers rounded to the nearest hundred.

d. It takes about 100 strawberries to make a kilogram. About how many kilograms of strawberries did he eat?

e. Did Mathy eat more berries in total in February or in April? How many more?

3. Four types of penguins are nesting on an island. Chinstrap penguins and macaroni penguins make up half the population on the island. There are 186,024 chinstrap penguins, which is half of the number of king penguins there are, and 80,053 more than the number of rockhopper penguins.

Your task is to find the number of each kind of penguin and the total number of penguins. Use blank paper for calculations.

Penguin Population	
Chinstrap	
King	
Macaroni	
Rockhopper	
Total	

a. With the information given in the problem, you can calculate the numbers of certain penguins. Which ones?

Do so. Write those numbers in the chart.

b. Use the information given in the problem *and* the numbers you have calculated *and* the bar model below to finish solving the problem.

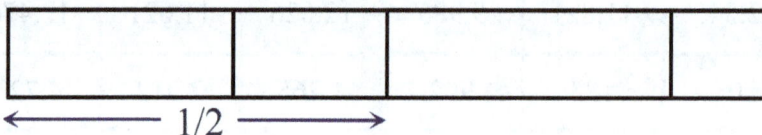

←———— 1/2 ————→

4. Solve the mystery numbers!

a. If you multiply me by 10, you would get the number 72,560. Who am I?

I am _____.

Mystery Number
38 2 11
47 101 99

b. I am six digits long. My hundred thousands digit is three times as large as my thousands digit. My tens digit is five times as large as my hundreds digit. My ones digit and my ten-thousands digit are the same, and add up to 8. My digit sum is 26. Who am I?

I am _____.

Multiplication in Word Problems

1. Write an equation with an unknown for each word problem, and then solve.

Problem:	Equation
a. One tricycle has three wheels, so ____ tricycles have 60 wheels.	$3 \times y = 60$ $y = $ _____
b. How many times more expensive is the jacket that costs $108 than one that costs $27?	
c. How many donkeys are there in a field if there are 52 feet?	
d. It costs $3,349 to buy 550 feet of fencing. How much fencing would three times that much money buy?	
e. A farmer is selling his sheep at $3 per pound. What is the total price for a young ewe that weighs 121 pounds *and* her 48-lb lamb?	
f. Phil earns $30 an hour. How much would he earn per week if he works 8 hours a day and 5 days a week?	

2. Solve.

During one week, an otter ate 48 oysters each day, for most of the days. But on the first day, it ate three more than that, and on the next to last day of the week, it ate six less than that. How many oysters did it eat that week?

3. Solve.

a. A tuft of grass in Africa grows 26 inches tall before being eaten by an antelope, after which it is three inches tall. It re-grows fully before being re-eaten by a zebra so it is three inches tall again. After this happens three more times, how much will the grass have grown and re-grown in total?

b. Martha buys a 9 × 17 ft shag rug with a zebra pattern to go in the living room, which measures 16 × 21 feet. How much of the living room floor is *not* carpeted, in square feet?

c. A store used to sell a certain blueberry-flavored energy bar for $3.50. At that price, they sold about 10 bars per day. Then they raised the price by one dollar, and the amount of sales dropped down to seven per day. In which scenario does the store take in more money?

4. Solve the mystery numbers! (All mystery numbers are less than 100.)

Mystery Number 38 2 31 11 99 47 01 9	**a.** I am between 30 and 60. The number one more than me is in the table of seven. The number one less than me is in the table of nine. I am _____.
b. The quotient of my digits is 3. I am divisible by 3, but I am over 36 and not in the standard multiplication table. I am _____.	**c.** I'm in the tables of seven and five, and the sum of my digits is more than the product of my digits. I am _____.

Multiplication and Estimation

Example. *Apple saplings cost $34 each. Georgia wants an apple orchard. How many saplings can she buy with $600?*

We can use estimation and mental math to get a rough idea of how many saplings she can buy. Let's round $34 to $30 so we can use mental math.

Now, 20 × $30 is $600, which means the answer will be less than 20 trees.
(Why? Because thee real price, $34, is *more* than $30.)

So, let's guess that it will be 18 trees, and multiply to find the cost (on the right):

So, 18 trees will cost $612. That means she cannot get 18 trees, but she *can* get 17. Those 17 will cost $612 − $34 = <u>$578</u>.

```
          3
        3 4
      × 1 8
      ------
      2 7 2
    +3 4 0
    ------
      6 1 2
```

1. Solve.

a. How many energy bars that cost $2.60 can Gloria buy with $25?

b. A set of 20 tennis balls costs $18.69. How many sets can Virginia get with $100? (Use estimation.)

How many balls will she get?

c. A tub of yogurt costs $4.76. How many of those can you get with $25?

d. A dining room chair costs $42.50. How many of those can you get with $200?

e. How many kilograms of flour can Jeanine get with $50 if one kilogram costs $5.65?

2. Solve.

a. An ostrich is covering 5 meters per stride. It runs a distance of 155 meters, going 55 km/h. How many strides was that?

b. A blue whale weighs 115,000 kg and an elephant weighs 5,000 kg. How many elephants would weigh more than the blue whale?

c. If an octopus has 280 suction cups on each of its tentacles, how many suction cups does it have in total?

d. Tickets for a science museum are $21.50 per person, and tickets for an aquarium are $24.95 for an adult and $14.95 for children. The Alcott family has two adults and three children.

Which place would be costlier for the family to visit? (Use estimation.)

How much is the exact total for the family to visit the cheaper place?

Mixed Word Problems 1

1. Solve.

a. Audrey needs to buy six pillows for her family. One kind of pillow costs $16.08 per pillow. Then there is a set of two pillows, costing $45.65 for the set.

Audrey decides to buy the more expensive pillows because they are better quality. How much is her total?

b. Jenna feeds her chickens 9 pounds of food a day. The food comes in 50-pound bags, and she has just opened a bag. In how many days will she need to open a new bag?

2. Write your own word problem to fit the given equation! You can give it to a family member or friend to solve.

a. $83 \times y = 332$

b. $20 - (5 \times 3) = x$

3. Camilla's horse Dandy needs new horseshoes every six weeks. Each of the four shoes costs $5.50. The farrier charges $44 for trimming and $120 for shoeing all four feet.

   a. How much does trimming and shoeing all four hooves cost (including the new shoes)?

   b. One shoe falls off while Dandy is in the pasture. If the farrier charges 1/4 the price to trim and shoe just one hoof, but the price of shoes remains the same, how much does it cost?

4. The Wagner family went on a trip to the beach 62 miles away, but they got lost on their way back. They were supposed to make a turn one mile down a loop, but they drove around the loop four times and only found the turn on the fifth round. When they got home, they had driven 140 miles (to and from the beach). How long was the loop?

Puzzle Corner   If a chicken lays an egg precisely every 25 hours, will it lay 30 eggs in a 30-day month or not? (Assume that the first egg is laid 25 hours into the month.)

If not, how many did it lay?
And if yes, how many spare hours did it have
(before it would have become the 31st day)?

Dealing With Time

1. Solve.

a. An airplane takes off at 4:17 PM for a flight of 3 hours and 20 minutes. When will it land?	**b.** It's 10:46 AM, and Olivia has been doing school work for one and a half hours. When did she start?
c. After a two-hour drive to the beach, the Elford family swam for two hours, ate lunch from 12:25 to 1:10, swam for another 80 minutes, and then drove home. How long was their trip?	**d.** Mom sets aside a time from 8:15 AM to 8:45 AM to clean some of the house every day. How long does she spend cleaning per week?
e. If Joanna puts $3 in her piggy bank every day for a year, how much will she have saved up by the end of the year?	**f.** If it takes 23 minutes to drive to a restaurant, at what time should Felix leave in order to arrive at 5:30?

g. Malcolm starts a painting on Wednesday and spends two and a half hours a day working on it. He finishes it on Saturday, after working fifteen minutes longer than on the rest of the days. How many hours and minutes did it take to complete the painting?

2. Larry has eight pastures for his sheep. He has them on each pasture for five days before switching them to the next one. Make a schedule for him so he knows on which days he needs to move them to a new pasture. You can use a calendar to help.

Larry's pasture rotation	
July 17 - July 21	Pasture 1
July 22 -	Pasture 2

a. How many days of **rest** (when the sheep are not there) does each pasture get? (They go back to Pasture 1 after they finish with Pasture 8.)

b. How many pastures would he need if he wanted each pasture to get seven weeks of rest?

c. He decides to split each pasture in half, so he has double as many pastures, but to move the sheep every three days. How many days of rest would each pasture have then?

Practice 24-Hour Time (Version 1)

These are two different flight routes going from Fresno to Stockholm. (Note: to keep things simpler, the flight schedules below do *not* take into account the time zone changes. Instead, all times are in the time zone of Fresno, California.)

Route 1	
16:21	Depart from Fresno
17:26	Arrive at San Francisco
	Layover
20:50	Depart from San Francisco
07:40	Arrive at Copenhagen
	Layover
10:45	Depart from Copenhagen
11:55	Arrive at Stockholm

Route 2	
09:47	Depart from Fresno
12:05	Arrive at Denver
	Layover
13:00	Depart from Denver
20:30	Arrive at Frankfurt
	Layover
02:05	Depart from Frankfurt
04:10	Arrive at Stockholm

1. How long is the flight from Fresno to San Francisco?

   What about the flight from San Francisco to Copenhagen?

2. Gustavo is taking Route 2. His watch says it is 11:12 PM.

   a. What time is it in 24-hour time?
   Where is he?

   b. How long will it be until he reaches Stockholm?

3. Rachel would prefer to be on a flight for the whole night so she can sleep without switching planes. Which route should she take?

4. Which route should Eric take if he wants to take the route that takes less time?

Practice 24-Hour Time (Version 2; more challenging)

Below you see two different flight routes going from Fresno to Stockholm.

Route 1	
16:21	Depart from Fresno
17:26	Arrive at San Francisco
	Layover
20:50	Depart from San Francisco
16:40	Arrive at Copenhagen
	Layover
19:45	Depart from Copenhagen
20:55	Arrive at Stockholm

Route 2	
09:47	Depart from Fresno
13:05	Arrive at Denver
	Layover
14:00	Depart from Denver
05:30	Arrive at Frankfurt
	Layover
11:05	Depart from Frankfurt
13:10	Arrive at Stockholm

Note these time zone differences:

- There is no time difference between Fresno and San Francisco

- Denver is 1 hour ahead of California

- Frankfurt is 8 hours ahead of Denver

- Copenhagen is 9 hours ahead of San Francisco

- There is no time difference between Copenhagen and Stockholm, nor between Frankfurt and Stockholm

1. How long is the flight from Fresno to San Francisco?

   What about the flight from San Francisco to Copenhagen?

2. Gustavo is taking Route 2. His phone is keeping track of the time zone changes and shows it is 1:10 PM.

   a. What time is it in 24-hour time?
   Where is he?

   b. How long will it be until he reaches Stockholm?

3. Rachel would prefer to be on a flight for the whole night so she can sleep without switching planes. Which route should she take?

4. Which route should Eric take if he wants to take the route that takes less time?

Units of Length

1. Solve. Write an equation or several to show your calculations for each problem.

a. How many crates that are 1 foot 7 inches tall each can you stack before you reach the ceiling of a shed, which is 8 feet high?

b. A mile is 5,280 feet. Grace walks 1,410 feet from her house to school every morning, then back. How many feet does she walk in a school week (five days)?

About how many miles is that? (Use rounded numbers.)

c. Judy collected some caterpillars from the forest and measured them. Their lengths were: 4 cm, 6 cm, 37 mm, 24 mm, and 18 mm.

How much longer is the longest caterpillar than the shortest?

If you lined them all up, how long a "train of caterpillars" would you get, in millimeters?

d. Charlie the dog runs through the house with muddy feet for four yards before being caught by his owner, who drags him three more yards plus two feet to the bathtub. How many feet did Charlie travel with muddy feet?

Mixed Word Problems 2

1. The line graph depicts the head of cattle per year at two different ranches.

Cattle Herd Size

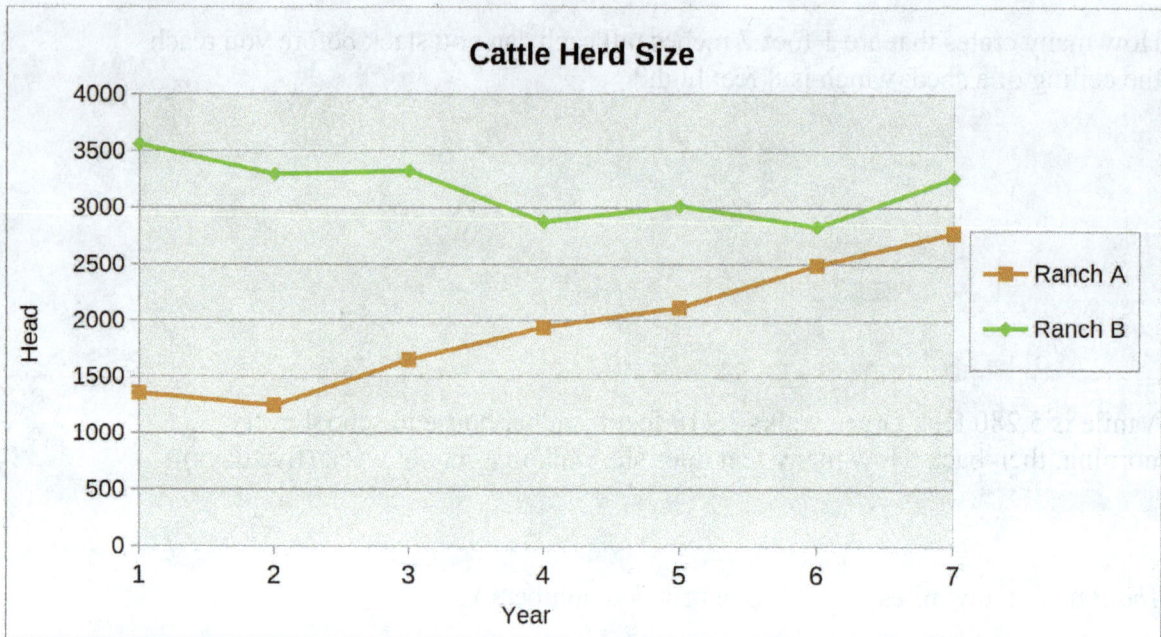

a. *About* how many more cattle did Ranch B have than Ranch A on the 4th year?

b. At what point was the amount of cattle at Ranch A double what it had been at its lowest?

c. When was the difference between the cattle on Ranch B and on Ranch A the greatest?

2. Tim and Bill are flying drones. Tim's drone is at 1,525 feet and Bill's drone is at 1,342 feet. During the next four minutes, the drones' heights are recorded as changing like this (in feet):

Tim's drone: −35, −65, +3, −21
Bill's drone: −32, +54, +64, +46

At what point(s) were the drones at the same height?

3. Solve.

a. Which watermelon costs less: a 16-pound watermelon priced at 12 cents a pound, or a 14-pound watermelon from a different market, priced at 14 cents a pound?

b. Kayla is running a 5-kilometer path in a forest. After every 800 meters, she takes a break and walks for 100 meters for 1 minute before running again. By the time she's finished the path, what distance will she have actually been *running*?

c. Oskar has $342 and wants to buy a camera that costs $1559. He earns $140 a day five days a week, but also spends $485 a week on various expenses. How long will it be before he can buy the camera?

Division Warm-Up

1. Write an equation for each situation. It might not always be division! Sometimes, several different equations can work. Then write what thing (what fact) you find out by your calculation.

Problem:	Equation:
a. A herd of deer has a total of 36 legs. *There are nine deer.*	$36 \div 4 =$ _____
b. Carlos divides each of the four apples into fourths.	
c. Sixty apples were packaged into 12 bags.	
d. Lee buys $20 worth of pineapples that cost $4 each.	
e. Hank the Great Dane knocks 13 of the 246 books off the bookcase.	
f. You buy four shirts for $5.50 each.	
g. Eight notebooks costs $24.	
h. 200 pencils are packaged evenly into four boxes.	
i. Jerry, Jason, and James share 12 squares of chocolate evenly.	

2. Find which calculation solves the problem. Then solve.

Pizzas cost $14 each. Six friends share the cost of three pizzas equally. How much did each pay?

$6 \div 3 \times 14 = x$ $x = 6 \times 14 \div 3$ $3 \times 14 \div 6 = x$

3. Solve. Watch out for the remainders!

a. Randy packages 100 apples into bags of 12 each.
How many full bags does he get?

How many apples are left?

b. Tamra takes on average four minutes to respond to each
customer e-mail. She starts answering emails at 9:08 AM
and finishes at 11:21 AM. About how many e-mails did
she answer? (Round.)

c. The granola at Bianca's health food store usually costs $24,
but it went over the expiration date, so she's selling it at 2/3
of its original price. What is it discounted by?

d. How long does a 50-pound bag of food last if a flock of
16 ducks eats four pounds a day? (The remainder does
not count as a "day".)

e. Of the tomato seeds Milo sowed, 29 sprouted. How many rows of tomato plants will he get
if he puts seven plants in a row?

And how many plants are in an incomplete row?

What about if he used rows of six?

An Expression for a Word Problem

Sometimes, writing a single calculation (or a single equation) can help you understand a word problem. In most cases, breaking the word problem down into multiple calculations, one for each step, might be easier. But we can also piece together a number sentence step by step and then solve it in one go.

Example. *On a big sale, Louise buys a set of six towels that cost $7 each and two pillows that each cost $12, then gets a half price discount. How much did she pay?*

It is asking for the total price of the towels and pillows Louise is buying. So, we can start by writing an expression for the price of the towels: 6×7.

She also buys two pillows for $12 each. Since it's in addition to the towels, we can add those two prices together. So, we have $6 \times 7 + 2 \times 12$ for the total cost.

Lastly, we see that she gets a half price discount, which means we have to halve the price. To do that, we put parentheses around the previous expression, and divide it by two: total cost $= (6 \times 7 + 2 \times 12) \div 2$.

Now all we have to do is solve it using the order of operations:

$$x = (\underbrace{(6 \times 7)}_{42} + \underbrace{(2 \times 12)}_{24}) \div 2$$

with $42 + 24 = 66$.

Louise paid $33.

---

1. Write a single expression (number sentence) for each word problem, and then solve.

a. Last year, Theodore bought nine packs of a dozen golf balls each. Since then, he lost 28 in a pond, 37 in the other pond, and four to unknown locations. How many golf balls does he have left?

b. Matthew bought four packages of resistors for a total of $28 and six packages of batteries for a total of $36. How much does one package of resistors plus one package of batteries cost?

2. Match each word problem to a matching equation, and solve.

a. Noah is low on money. He earned $36 over the last couple days, but before that he only had $4. Today he spent $7 on snacks. How much does he have now?

i. $(36 - 4 \times 7) \div 4 =$ _____

b. Luna's four 2-month-old puppies weigh a combined total of 36 pounds, with each weighing the same amount. A few weeks later, they've each gained seven pounds. How much does one of them weigh now?

ii. $(7 + 4) \times 7 + 36 =$ _____

c. Mr. Forrest buys a drawing pad for $7 and a set of pencils for $4 for each of his seven art students. He buys an additional $36 worth of charcoal, erasers, and other supplies. How much did he spend in total?

iii. $36 \div 4 + 7 =$ _____

d. Harry had $36. He then buys a $1 drink every day for four weeks. In that time, he also earns another $36. How much does he have now?

iv. $36 - 7 \times 4 + 36 =$ _____

e. Java the duck has laid 36 eggs over the last two months. Then she lays an egg every day for the next four weeks. How many has she laid now?

v. $4 + 36 - 7 =$ _____

f. Jordan is building a tree house. He plans for it to be 4×7 feet, but he has to add a small rectangular section to one of the sides so it will fit in the tree. He ends up with an area of 36 ft². One side of the extra section was 4 feet long. How long was the other side?

vi. $36 + 7 \times 4 =$ _____

Division Practice

1. Makayla's hummingbird feeder holds 750 grams of sugar water, and she has to refill it every three days. How much are the hummingbirds drinking in a day?

2. Dad paid $89 in total for an $11 umbrella plus six identical towels. How much did one towel cost?

3. A rancher owns 8,049 acres of land. He has 3 acres per cow. How many cows does he have?

4. Last year, Clarissa's eight geese laid 256 eggs in total. This year, each goose laid four eggs *less* on average. How many eggs did she get this year?

5. Nancy goes to soccer practices that usually last two hours but sometimes three. She figured that last year, she had spent a total of 141 hours in those practices. If 48 of those practices had lasted two hours, how many of them had lasted three hours?

6. Elijah bought two jars of sauerkraut and five jars of jam for a total of $94.55. At home he wanted to check the prices of each item but couldn't find his receipt. The sauerkraut jar had a sticker saying it cost $18.90. How much did each jar of jam cost?

7. For unknown reasons, Mathy the Mammoth has ordered six trucks full of blueberries to be dumped in his backyard, with a total of 7,830 kg of blueberries. Each truck would haul the same amount. However, the second truck accidentally dumps a third of its load at Mathy's friend's house, before bringing the rest to Mathy. How many kilograms of blueberries did Mathy actually get?

8. Write your own word problem to fit the given equation! You can give it to a family member or friend to solve.

a. $123 \div 6 = x$

b. $48 \div 12 + 3 = x$

Hayley uses ninety-five 50-lb bags of chicken food in a year. If each chicken eats about half a pound a day, how many chickens does she have?

40

Units of Volume

1. Solve.

a. The Greenwood family buys a 4-liter jug of orange juice. If two of the family members drink one 300-mL glass of orange juice a day, and the other three family members each drink one 250-mL glass of orange juice a day, how long will it last?

b. Jenna fills a 15-gallon tub 4/5 of the way full with water for her ducks. How many quarts of water is that?

c. Cassie eats half a pint of yogurt for breakfast every day.

In how many days will she have eaten a quart of yogurt?

How long would a gallon of yogurt last her?

If yogurt costs $5.35 per quart, how much will it cost her to buy enough yogurt for at least one week?

d. Sherry is using orange juice concentrate to make juice by adding 3 oz of water to 1 oz of concentrate. She wants 4 cups of juice. How many ounces of orange juice concentrate and how many ounces of water should she use?

Finding the Average

Example. For a week, Leon tracks how long he spends drawing each day, in minutes. The results are: 52, 29, 0, 88, 23, 109, 0. On average, how long did he spend drawing per day?

To find the average, first find the total number of minutes he spent drawing:

$$52 + 29 + 0 + 88 + 23 + 109 + 0 = 301$$

Then, divide that by however many numbers you added (in this case, seven), using long division:

$$301 \div 7 = 43$$

On average, Leon spent 43 minutes drawing per day.

1. Solve the problems.

a. Manuel drove 378 miles in nine hours. What was his average speed?

b. Melanie drives with an average speed of 49 miles per hour for six hours. What distance did she drive?

c. Watermelon plant A made three watermelons that weighed 22, 23, and 18 pounds.
The watermelons of plant B weighed 23, 20, 16, and 21 pounds.
Which plant made bigger watermelons on average, and how much bigger?

2. Last year, Kelsey's nine mango trees produced 4,032 mangoes. This year, each tree produced on average seven more mangoes than it did last year. How many mangoes did all the trees produce this year?

3. Jasper is studying wandering albatrosses. He measures the wingspan of five different birds and gets these results, in inches: 128, 98, 105, 114, 120.

What is the average wingspan of the wandering albatrosses, in feet and inches?

Then he measures one more bird with a wingspan of 137 inches. Now what is the average?

Puzzle Corner Lemon ginger tea costs $5.40 for a box of 18 teabags. Sasha uses one teabag a day. How much will she spend on tea in a month (30 days)?

Units of Weight

1. Solve.

a. Blueberries come in packages of 120 grams each. Mathy needs more blueberries than that, so he buys 50 boxes of blueberries. How many kilograms of blueberries is that?

b. Six newborn Labrador puppies have the following weights: 482 g, 493 g, 482 g, 471 g, 488 g, 486 g.

What is their average weight?

At eight weeks, the largest puppy now weighs 6 kg 45 g. How much did it gain since birth?

c. A lamb is born weighing 9 pounds 4 ounces.
If it gains about two and half pounds a week,
how long will it be until it is 50 pounds?
If you find this difficult, see a hint at the bottom of the page.

d. (challenge) Caleb usually uses four packages of cream cheese in his cheesecake recipe, which calls for two pounds of cream cheese. This time, his cream cheese packages are 12 oz each, which is larger than normal. How many packages of cream cheese does he need now?

Will there be any cream cheese left over? How much?

Hint: 2 ½ is a number that works well with 10. In how many weeks does the lamb gain 10 lb?

Divisibility and Factors

1. Find the mystery numbers! All numbers are less than 100.

a. This number is divisible by 7 and 5, and is more than 50.

b. This number is divisible by 2, 3, and 4, and is between 90 and 100.

c. List all the primes between 90 and 100.

d. This number has exactly nine factors and is between 30 and 40.

e. This number has exactly 12 factors and is between 80 and 90.

f. This number has exactly three factors and is between 40 and 70.

2. Claudia said, "I have a secret number that is less than 100, and is only divisible by 3 and 17." Jeremy said, "That can't be true. It's divisible by two other numbers also."

What are those two other numbers?

3. Luis said, "I found two primes that are only two numbers apart, and they're both between 30 and 40." Alejandro said, "That's nothing. I found *three* primes that are like that; they're consecutive odd numbers."

One of the boys is not correct. Who? And why? (Explain.)

4. Randy said, "This number is more than 60, less than 100, and it is divisible by 4 and 6 but not by 8." What is Randy's number?

5. Find all the numbers between 20 and 30 that have exactly four factors.

6. Eleanor, "My number is less than 50, has exactly five factors, and is divisible by 3."
   Jessica, "That cannot be. You must have made an error. A number with exactly five factors that is less than 50 does exist but it is not divisible by 3."

Who is correct?

Puzzle Corner   Find all the two-digit numbers that have exactly three factors.

Area and Perimeter

1. This shows the floor plan of a small house.

   a. What is the area of the whole house?

   b. What is the perimeter of Room A?

   c. What is the area of Room A?

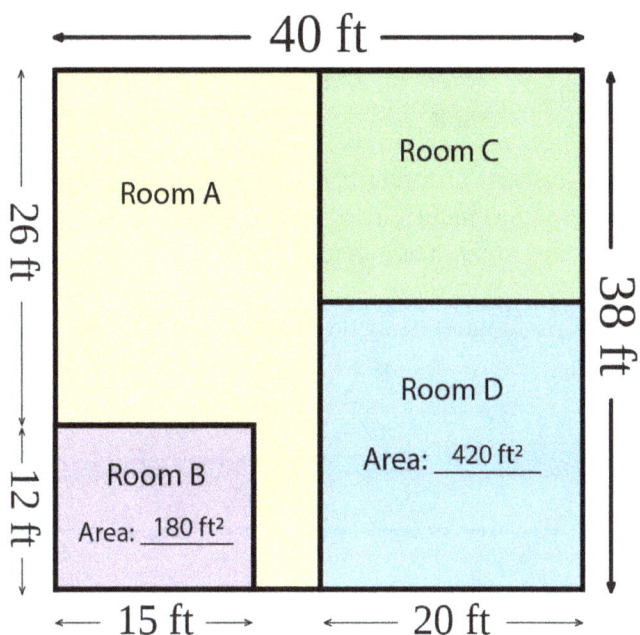

40 ft

26 ft

12 ft

38 ft

Room A

Room C

Room D
Area: ___420 ft²___

Room B
Area: ___180 ft²___

15 ft   20 ft

2. Scott is tiling a 15 ft x 17 ft room. The tiles he is using are 12 x 12 inches. How many tiles will he need?

3. Ivy's porch has an area of 21 m². The side facing the house is 7 meters. Ivy can't enjoy sitting on the hammock on her porch because of mosquitoes, so she wants to put screen netting all around. How many meters of screen will she need?

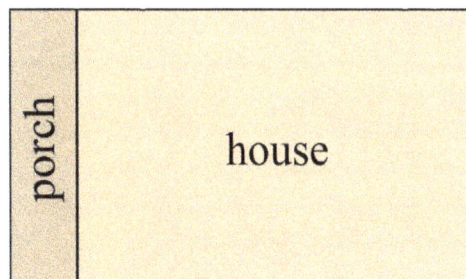

porch

house

4. (challenge) Ariel has fenced in an area of 30 × 12 feet to use for a garden. She now needs to decide where in that garden to put the actual garden beds. The garden beds will have walkways between them that need to be at least 18 inches wide.

The garden beds should be either (a.) no more than 3 feet wide IF there is a walkway on both sides, or (b.) no more than 18 inches wide IF the bed is against the fence (so she can reach across it without stepping in it.)

Suggest two different ways she can arrange beds in her garden! Each square in the grid below measures 6 inches by 6 inches. You can make garden beds or rows in any rectangular shape. Be sure to include a gate anywhere you want.

Then calculate: how much actual garden area does each plan give her?

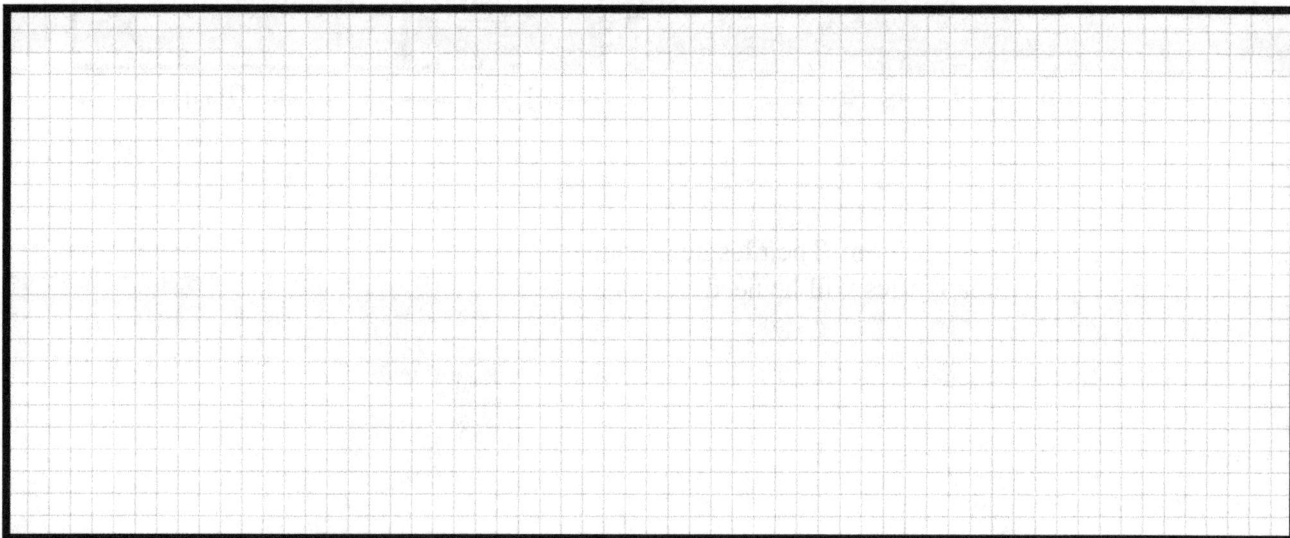

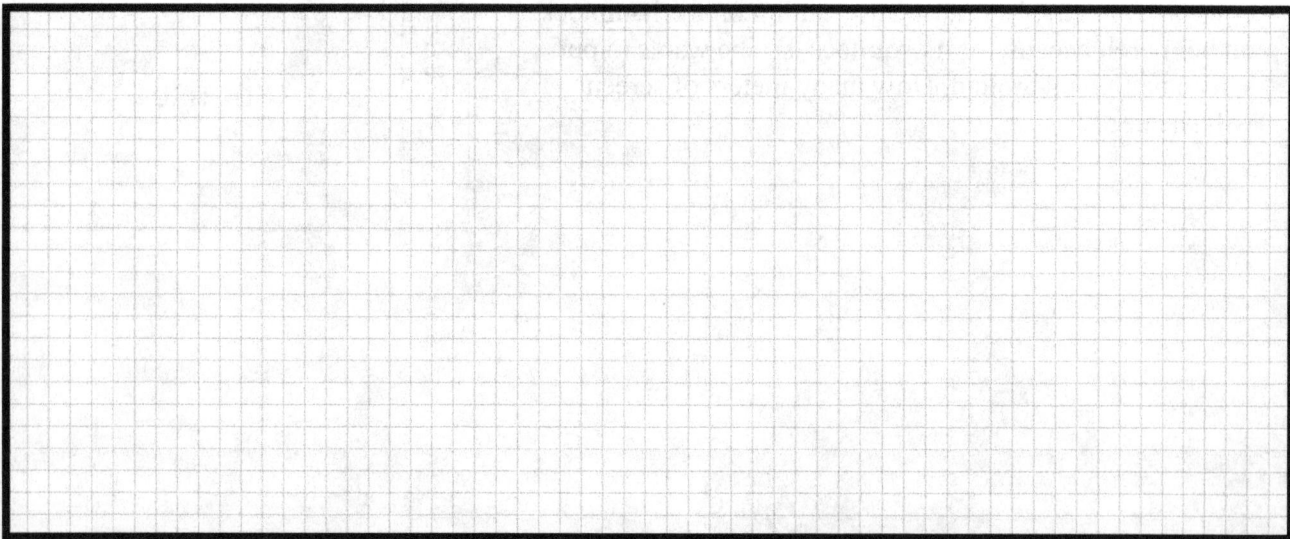

Angles and Shapes

1. Solve.

a. Hailey drew a rectangle and then drew a diagonal in it, from point A to point C. Then she measured angle CAD and said, "This angle measures 117°."

How can you *easily* tell (without measuring) that she didn't measure it correctly?

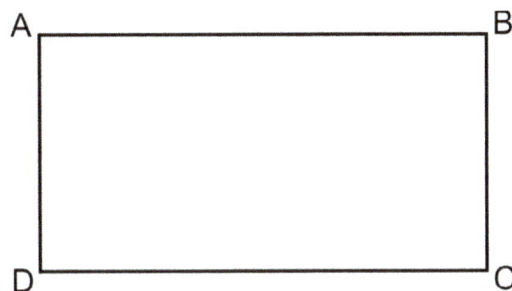

What error did Hailey probably make?

```
A                                    B



D                                    C
```

b. What is the angle sum of these three angles? One angle is a right angle. Another measures 1/3 of the first. The last angle measures 1/2 of the second angle.

c. Sarah wrote a riddle about three angles. "These three angles, x, y, and z, share a vertex and sum up to 360°. The first one is an acute angle. Its angle measure is the only number between 60 and 70 that is divisible by eight. The other two are obtuse and have the same measure."

What are the measures of the three angles?

2. Henry said, "I started out by drawing an angle. Then I drew a line to get a triangle. So, since the angle I drew first was acute, that is why it is an acute triangle."

   Is he correct? Explain.

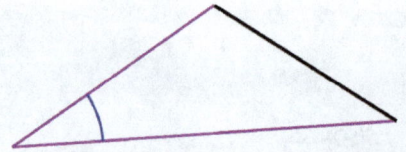

3. Here you see a quadrilateral. The line drawn from one vertex to the opposite vertex is called a **diagonal**. It divides each shape into two triangles.
   Draw a quadrilateral, and a diagonal in it, so that the two triangles formed are....

   a. obtuse triangles

   b. right triangles

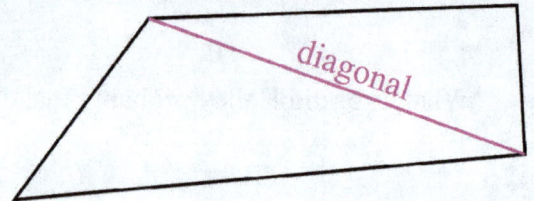

4. Diana described what she drew:
   "First I drew two points, A and B, and a line through them. Then I drew point C above that line. Then I drew a line through C that was perpendicular to the first line I drew. Lastly I connected points C and B with a line segment."

   What shape did Diana get?

Fraction Fun

Example. There are 10 cups of oats in a box of oats. Oatmeal calls for 1 3/4 cups of oats. How many times can Mom make oatmeal using one box of oats?

Let's draw a picture to help us solve this. We will draw 10 individual squares, symbolizing the 10 cups, divided into fourths.

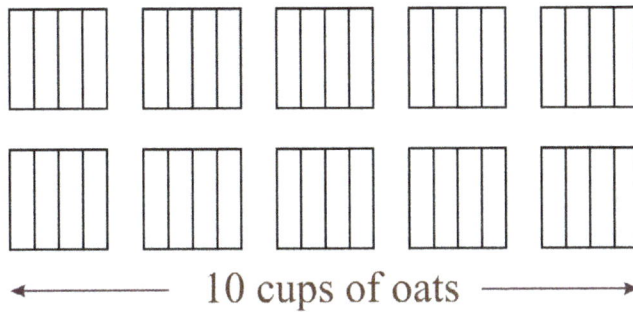

← ——— 10 cups of oats ———→

Now you can color in 1 3/4 cups of oats, repeatedly, in the big picture of 10 cups of oats, and keep track how many times you do it, to find out how many batches of oatmeal Mom can make out of that box of oats.

1. Solve. Draw pictures to help you!

a. Solve the problem in the above example.

b. Wendy made bread and cookies on the same day. The bread took 5 ½ cups of flour and the cookies took 2 ½ cups of flour. How much flour did she use in total?

c. Ronald ran 6 ¾ laps of a running track and Julie ran 2 ¼ laps. How many fewer laps did Julie run?

2. Solve. Draw pictures to help you!

a. A recipe calls for 1 1/3 cups of butter, but you are making a triple batch. How much butter will you need?

b. Brayden takes 1 ¼ hours to go bicycling. If he started at 4:10 PM, at what time did he finish?

c. At two months of age, a tomato plant was 1/2 meter tall. Two months later it was five times as tall! How tall is it now?

d. Three cats shared five fish of the same size unevenly. One cat ate 2 1/5 fish and the second ate 1 3/5 fish. How much fish did the third cat get?

e. A deer walked 3/4 of a mile, stopped to graze, and walked 2/4 of a mile more in the same direction. Then it walked all the way back. How many miles did it travel in total?

Puzzle Corner If Cheyenne walks at 4 miles per hour, how many miles will she travel in 1/4 of an hour?

How long would it take her to walk 1/3 of a mile?

Fractions and Division

Example. The Andrews family drives 171 km to the beach. When they're 4/9 of the way there, they stop for ice cream. How much farther do they have to go?

The fraction 4/9 indicates that this total of 171 km is divided into **9 parts** and that the family has driven **four** of those parts:

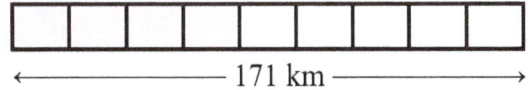

←——————— 171 km ———————→

To start, we can use long division to find what is 1/9 of 171. (Then, what is the next step?)

1. Solve.

a. Finish solving the problem in the example above.

b. On the way back, the Andrews family stops for gas 60 km from home. *About* what fraction of the way home are they? Use a rounded number and estimate.

c. Of the 12 birds Randy sees, three are robins, and a third are finches.

What fraction of the birds were robins?

How many finches were there?

How many birds were NOT robins or finches?

d. A cheesecake weighs 850 g. It is cut into ten slices.
How much do three slices of it weigh?

2. Solve.

a. Lily bought a smoothie for $12.60 and a cheaper smoothie which cost 1/3 as much. What was her total cost?

b. Fletcher sawed off one-fourth of a 16-foot plank of wood. Then, he sawed the *rest* of the plank of wood into three equal pieces. How long is the first piece he sawed off?

What about each of the other three pieces?

c. Emma starts a lemonade stand, selling lemonade at $2 per cup. She sells 52 cups the first day, and then spends 1/5 of the money buying more lemons and sugar for the next day. How much money does she have left after that?

d. There are 2 hours and 20 minutes left before Susan eats supper. From now until supper, Susan spends 5/7 of her time working, and then plays a game the rest of the time. How much more time did she spend working than playing, in minutes?

Decimals

1. Solve.

a. Which is shorter: 1.5 feet or 1 foot 7 inches?

b. A playground measures 18.2 meters by 21.5 meters (length and width). What is its perimeter?

c. Gerald buys three chocolate bars for $5.29 each and pays with $20. What is his change?

d. Bridget walked to the grocery store and back, a total of 1,400 meters. Philippa went for a 2.4-kilometer walk for exercise. Who walked more, and how many more meters?

e. Which one has a larger perimeter: a 5.74 m by 2.8 m rectangle OR a square with 4.3-meter sides? How much larger?

f. A squirrel is moving a stash of nuts from one tree to another. The trees are 18.6 meters apart. The squirrel has made two trips between the trees so far (one round trip). How many more trips will it make before it has run 100 meters in total?

2. Solve.

a. Aurora spends the morning tending her garden and harvesting vegetables. At home, she weighed what she had picked: 0.85 lb radishes, 0.62 lb greens, 5.2 lb pumpkins, and 3.9 lb tomatoes. What is the total weight of her vegetables?

b. Daniel lives 0.6 miles from the nearest grocery store. Harold has to travel 1/2 mile to reach the same store.

Who has to go farther? How much farther?

Let's say they both bicycle to the grocery store and back every Tuesday and every Friday. What is the total distance each one of them bicycles on these trips?

c. A bread recipe for one loaf calls for 400 grams of flour. Tammy weighs the flour she has left in the bag, and gets 1.82 kg. How many loaves of bread can she make with that amount of flour?

d. Twin lambs weigh 8.6 pounds and 7.8 pounds. If they each gain 5.2 pounds a week for three weeks, what will be their *combined* weight at that time?

Puzzle Corner

There were 1.5 gallons of ice cream in the freezer. Then Dad ate some at lunch and supper, and Kirsten and Monica each ate 1.5 cups at both meals. Now there is 1 gallon left. How much did Dad eat?

56

Mixed Word Problems 3

1. Solve.

a. Zoey is planning to walk 1,530 meters around the block every other day, starting on Monday. How far will she have walked by Sunday?

b. If a pack of four identical shirts costs $23.60 and a pack of eight such shirts costs $47.20, how much would you expect a pack of six such shirts to cost?

What about a pack of five?

c. Jonathan is raising rabbits for meat. Last year, he started out with 14 rabbits. Now, his rabbit population has grown five-fold from what he started. He expects that next year, it will be triple what it is now.

How many rabbits does Jonathan think he will have next year?

d. Ricky spent 1/5 of his savings on a magnifying glass. Now he has $20 left. How much did the magnifying glass cost?

2. Spencer wants to calculate the area of a certain pasture, which is in the shape of a square. Using a tape measure, he finds that the distance from the first corner to the gate is 26 meters. The gate itself is 3 m wide. These 29 meters are exactly 1/3 of the length of that side.
(Hint: draw a sketch.)

Can he already figure out what the area is?

If not, what else does he need to do?

If yes, what is the area?

3. Spencer is also planning to make another, rectangular pasture using 251 meters of fencing he already has. This pasture will share one side with the previous pasture, to save on fencing. What will be the length and width of this new pasture?
(Hint: draw a sketch.)

Answer Key

Identify the Operation, pp. 7-8

Page 7

1. a. $123 + 94 = $ ____ or ____ $- 123 = 94$. Answer: He had $217 before.
   b. $48 \div 8 = M$. Answer: Each mug cost $6.
   c. $6 \times S + 23 = 41$ or $(41 - 23) \div 6 = S$. Each snack bar cost $3.

Page 8

2. a. $260 + 128 + \underline{342} = 730$ or $730 - 260 - 128 = \underline{342}$. The third number was $\underline{342}$.
   b. $239 \times 4 = \underline{956}$. He drove $\underline{956}$ miles.
   c. $4 \times 9 = \underline{36}$. Elaine needs $\underline{36}$ apples.
   d. $131 - 119 = \underline{12}$. Their heights are $\underline{12}$ cm apart.
   e. $\$500 - \$249 = \underline{\$251}$. Denise spent $\underline{\$251}$.
   f. $7 \times 3 = \underline{21}$. $41 - 21 = \underline{20}$. $20 \div 2 = \underline{10}$. There are $\underline{10}$ bikes.

Addition and Subtraction, pp. 9-10

Page 9

1. a. $197 + x = 204$. $x = \underline{7}$.
   b. $x - 15 = 14$. $x = \underline{29}$.
   c. $x - 8.50 - 8.50 = 3$. $x = \underline{\$20}$.

2. a. Notice that the 60 minutes is not information needed to calculate the total distance.
      $22 + 25 + 18 = \underline{65}$. The total distance was $\underline{65 \text{ km}}$.
   b. $412 - 196 + 83 = \underline{299}$. Perry earned $\underline{\$299}$.
   c. $3 \times 30 + 60 + 45 = \underline{195}$. He spends $\underline{195}$ minutes exercising.
      $\underline{3} \times 60 = 180$. $195 - 180 = \underline{15}$. That is $\underline{3}$ hours and $\underline{15}$ minutes.

Page 10

3. a. The tallest pine tree was planted the first year, so there are three smaller pine trees.
      $3 \times 22 = 66$. $116 - 66 = \underline{50}$. The shortest tree is $\underline{50}$ inches tall.
   b. There are two ways to calculate this:
      $2 \times 237 = 474$. $2 \times 474 = \underline{948}$ or $2 \times 2 = 4$. $4 \times 237 = \underline{948}$. The flock has $\underline{948}$ geese now.

4. a. ii. $x - 48 = 123$
   b. iii. $123 - x = 48$
   c. i. $123 - 48 = x$

5. Answers will vary. Check the student's work. For example:
   Mathy is eating blueberries. He eats nine blueberries per bite and has taken 17 bites.
   How many blueberries has he eaten?
   $9 \times 17 = \underline{153}$. He has eaten $\underline{153}$ blueberries.

What Is It Asking?, pp. 11-12

Page 11

1. a. First find out: How many boys are there? $98 - 43 = \underline{55}$.
      Then, $55 - 43 = \underline{12}$. There are $\underline{12}$ more boys than girls.

   b. First find out: What is the total amount Cash needs to buy the laptop and the mouse? $989 + 9 = \underline{998}$.
      Then, $998 - 821 = \underline{177}$. He needs to earn $\underline{\$177}$ more.

What Is It Asking?, cont.

Page 12

2. a. First find out: <u>What was the total for the three bars?</u> $3 \times 3.75 =$ <u>11.25</u> or $3.75 + 3.75 + 3.75 =$ <u>11.25</u>.
Then, $20 - 11.25 =$ <u>8.75</u>. The change was correct.

   b. First find out: <u>How much time did she work on the nest on each of the next six days?</u> $6 \div 3 =$ <u>2</u>.
Then, $6 \times 2 + 6 =$ <u>18</u>. She took <u>18</u> hours to build her nest.

   c. First find out: <u>How many miles did they drive from home to the ski track?</u> $84 - 29 =$ <u>55</u>.
Then, $55 - 29 =$ <u>26</u>. They have <u>26</u> more miles to drive.

3. a. We are not told how many chicks the second chicken had, so cannot calculate the total number
of chickens and chicks.
   b. This one is a little bit of a trick question, as we are already given the answer, rather than being given
each of their heights and having to calculate that there is a difference of 15 cm.

Using Bar Models, pp. 13-14

Page 13

1.

a. The number of horses that are not chestnut are
$21 + 13 + 17 = 51$, or a little over half of the total.
There are $98 - 51 =$ <u>47</u> chestnut horses.

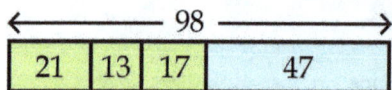

98
21	13	17	47

b. The number of books that have arrived are
$650 + 650 + 650 = 1,950$, or less than half of the
total. $5,000 - 1,950 =$ <u>3,050</u> books have yet to arrive.

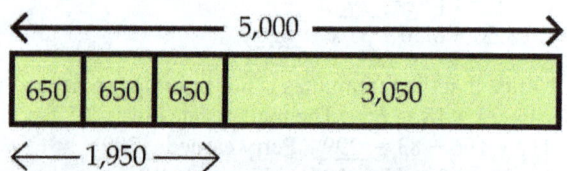

5,000
650	650	650	3,050

1,950

c. He has eaten $5 + 14 = 19$ kg of berries so far. Since
this is half of his breakfast, the total amount he is
eating is $19 + 19 =$ <u>38</u> kg.

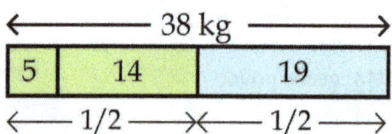

38 kg
5	14	19

1/2 — 1/2

d. Half of 48 is 24. $24 - 12 - 8 =$ <u>4</u> chickens
are gold.

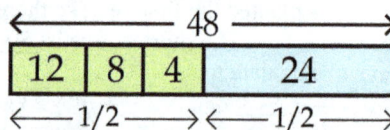

48
12	8	4	24

1/2 — 1/2

Page 14

2.

The problem is asking for the total of the distance to the gas station, plus the distance traveled since the gas station — the
first two blocks on the bar model. At the gas station, they had traveled $572 - 321 = 251$ miles. Now, they have gone
another 93 miles, so they have traveled $251 + 93 =$ <u>344</u> miles.

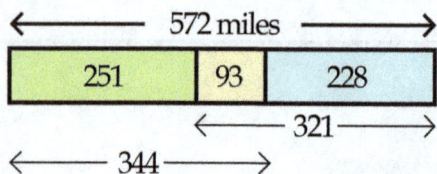

572 miles
251	93	228

321

344

3.

a. Dallas initially spent 584 − 465 = $ 119 to buy the puppy. Then, with the money he put aside, he is spending 119 + 120 = $ 239 on Jet.

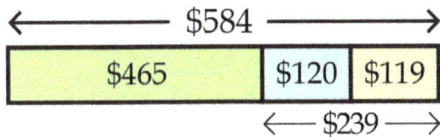

b. She paid equivalent to 1/2 + 1/2 + 3 = 4 months at the normal price. So the normal price of the program for one month is $44 ÷ 4 = $ 11 .

Two bar model options:

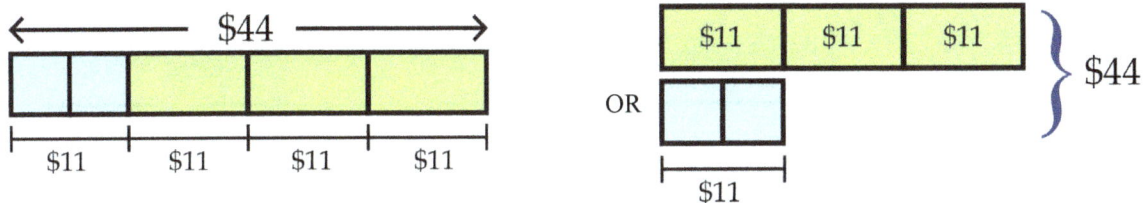

c. After making the first two necklaces, she had 400 − 153 − 96 = 151 *or* 400 − (153 + 96) = 151 beads. So she used 151 − 51 = 100 bead on the third necklace.

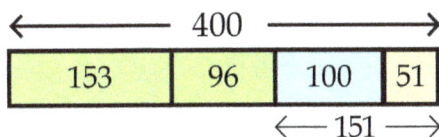

Money and Discounts, pp. 15-16

1.

a. We can write the equation: $x - 19 - 25 = 41$. This means that if we add the amounts that Griffin has spent, and what he has left, we can find that he had $19 + $25 + $41 = _$85_ originally.	b. $80 - x - x = 28$. To find how much Helen spent, subtract $80 − $28 = $52. One paint set cost half of $52, or _$26_.
c. $63.55 − $7.90 = $55.65 . It costs $55.65 now.	d. The three umbrellas cost 3 × $11.69 = $35.07. 2 × $20 = $40, so her change is $40 − $35.07 = $4.93.
e. Dad spent a total of $531 − $327 = $204. Of that, he spent $204 − $172 = $32 on flyswatters. So he bought 32 ÷ 4 = _8_ flyswatters.	f. The flashlights at the discounted price were $36 ÷ 3 = $12. So, the discount per flashlight was $14 − $12 = $ 2 .

2. a. Six shirts for $48 is 48 ÷ 6 = $8 per shirt. Four shirts for $36 is $36 ÷ 4 = $9 per shirt.
   So the <u>set of six</u> has cheaper shirts.

   b. First, round $3,269.95 up to the next hundred: $3,300. $3,300 ÷ $100 = <u>33</u>. So, thirty-three $100 bills will be needed. Her change will be $3,300 − $3,269.95 = <u>$30.05</u>.

   c. Corey's total so far is $99 + $53.68 = $152.68, so he needs to spend at least $200 − $152.68 = $47.32 more. Since the shovels cost almost $10 each, we might guess that five shovels would cost at least $47.32. Let's check: 5 × $9.50 = $47.50. (Or, you might notice that since each shovel costs 50 cents short of $10, then five shovels will cost $2.50 short of $50.) Yes, he needs to buy <u>five</u> shovels. His initial total will be $152.68 + $47.50 = $200.18. After the discount, his total will be $200.18 − $50 = <u>$150.18</u>.

Puzzle Corner: First, add the $2 discount to the $11, to determine how many pounds you could buy including the discount. $11 + $2 = $13. Then determine how many times 1.50 will go into 13. 10 × 1.50 = 15.00, which is too much. 15.00 − 1.50 = 13.50, which is still just a little too high. <u>8</u> × $1.50 = $12.00. Then take off the discount: $12.00 − $2.00 = $10.00. You can buy <u>8</u> whole pounds of potatoes with $11, after the discount.

Going Hiking, pp. 17-18

1. Estimate: From point A to point B is 2,503 ≈ 2,500 ft, and from point B to point E is 1,051 ≈ 1,000 ft.
   2,500 + 1,000 + 1,000 + 2,500 = <u>7,000 ft</u>.
   The exact distance is 2,503 + 1,051 + 1,051 + 2,503 = <u>7,108</u> ft.

2. You can see that it will be shortest to either go to point G, point E, and exit F, or to go point I and then exit J. You can first round to the nearest hundred and estimate which distance is shorter. The exact distance to exit F is: (3,061 − 1,305) + (4,557 − 2,182 − 880) + (2,989 − 1,051) = 1,756 + 1,495 + 1,938 = 5,189 ft. The distance to exit J is: 1,305 + (3,865 − 487) = 1,305 + 3,378 = 4,683 ft.
   The shortest route is <u>to point I and then exit J</u>. It is <u>4,683 ft</u>.

3. They started at F, and went through points E, G, I, H, and C, to D.

4. The opposite path is 3,865 + (4,557 − 2,182) + (2,989 − 1,051) = 3,865 + 2,375 + 1,938 = 8,178 ft.
   The opposite path is 8,178 − 7,252 = <u>926 ft</u> longer than Shadow Trail.

5. First find out: <u>what the approximate total distance is around the loop.</u>
   Rounding to the nearest hundred, one loop is about 8,200 + 7,300 = 15,500 ft. Since one mile is 5,280 feet, then 15,500 ft, or one loop, is *about* 3 miles. At 6 miles per hour, Jasper can go 12 miles in two hours. This means Ellie can ride Jasper around the loop <u>four times</u> in two hours.

   (To check using exact numbers, we get: 8,178 ft + 7,252 ft = 15,430 ft. Three miles is 3 × 5,280 ft = 15,840 ft. So, one loop is slightly less than three miles, and yes, she can ride four loops in less than two hours.)

6. Answers will vary. Please check the student's work.
   For example: ABEGIJ. 2,503 + 1,051 + (4,557 − 880 − 2,182) + 1,305 + (3,865 − 487) =
   2,503 + 1,051 + 1,495 + 1,305 + 3,378 = <u>9,732 ft</u>.

Thousands Upon Thousands, pp. 19-20

Page 19

1. a. Rounding to the nearest thousand: 29,000 ft − 20,000 ft = _9,000_ ft. Mount Everest is approximately _9,000_ ft taller than Mount McKinley.
   b. To make it easily divisible by 12, round $58,313 to _$60,000_.
      She earned about 60,000 ÷ 12 = $ _5,000_ each month. Since 58,313 ÷ 12 ≈ $4,859.42, this is a reasonable estimate.
   c. $3,198 is about $ _3,000_.
      He would earn about 12 × $3,000 = $ _36,000_ in a year.

2. a. Mathy ate 33,256 + 9,985 + 902 = _44,143_ berries in April.

   b. In June, Mathy ate about 3,000 strawberries and 12,000 blackberries.
      That makes a ratio of 3,000 : 12,000 = 3:12 = 1:4, or 1/4. It works out the same if rounding to the nearest hundred, but isn't quite as simple to calculate.

   c. He didn't eat any strawberries in Jan or Feb, so he ate about
      100 + 900 + 1,300 + 2,900 + 4,700 = 9,900 strawberries.

   d. He ate about 9,900 ÷ 100 = _99_ kg of strawberries.

   e. Mathy ate more berries in _April_. We have already calculated that he ate 44,143 berries in April.
      So he ate 44,143 − (29,910 + 12,545) = _1,688_ more berries in April than in February.

Page 20

3. a. There are 186,024 chinstrap penguins. Since that number is half the number of king penguins, there are
      186,024 × 2 = _372,048_ king penguins. There are 186,024 − 80,053 = _105,971_ rockhopper penguins.

   b. Since chinstrap and macaroni penguins make up half the total population, king and rockhopper penguins
      make up the other half. Write the equation: 186,024 + x = 372,048 + 105,971, from which 186,024 + x = 478,019.
      There are 478,019 − 186,024 = _291,995_ macaroni penguins. Add the numbers of each kind of penguin
      for the total.

Mac.	Chinst. 186,024	King 2 × 186,024	Rock. 105,971

←——— 1/2 ———→

Penguin Population	
Chinstrap	186,024
King	372,048
Macaroni	291,995
Rockhopper	105,971
Total	956,038

4. a. The number multiplied by 10 would have a zero tagged to the end: 72,56_0_.
      To find the number, remove the zero: _7,256_ × 10 = 72,560.

   b. Start with the tens digit and hundreds digit. The tens digit can't be 10, so it has to be 5, and the hundreds digit 1,
      so the number is formed as _ _ _,15_. Since the ones and ten-thousands digits are the same and equal 8, they are
      each half of 8, or 4. So, the number is: _4_,154. The digit sum so far is 4 + 1 + 5 + 4 = 14.
      That means the remaining two digits will add up to 26 − 14 = 12. 9 + 3 = 12 and 9 is three times as large as 3, so
      the full number is: _943,154_.

Multiplication in Word Problems, pp. 21-22

1. Student equations will vary; check the student's equation.
   It can be helpful to think of multiplication and division families.
   a. $3 \times y = 60$ $(60 \div 3 = y)$
      $y = \underline{20}$ $\underline{20}$ tricycles have 60 wheels.

   b. $27 \times m = 108$
      $m = \underline{4}$ The jacket that costs $108 is $\underline{4}$ times more expensive.

   c. $4 \times d = 52$
      $d = \underline{13}$ There are $\underline{13}$ donkeys in the field.

   d. Three times as much money would buy three times as much fencing.
      $3 \times 550 = f$
      $f = \underline{1,650}$ It would buy $\underline{1,650}$ feet of fencing.

   e. $3 \times (121 + 48) = t$
      $3 \times 169 = t$
      $t = \underline{507}$ The total price for the ewe and the lamb is $ \underline{507}$.

   f. $30 \times 5 \times 8 = w$
      $w = \underline{1,200}$ He would earn $ \underline{1,200}$.

2. It ate $7 \times 48 + 3 - 6 = \underline{333}$ oysters that week.

3. a. The grass first grows to 26 inches. Then it grows from 3 inches to 26 inches once, plus three more times.
      The grass has grown $26 + 4 \times (26 - 3) = 26 + 4 \times 23 = 26 + 92 = \underline{118}$ inches in total.

   b. The living room floor is 16 ft $\times$ 21 ft $= \underline{336 \text{ ft}^2}$. The carpet is 9 ft $\times$ 17 ft $= \underline{153 \text{ ft}^2}$.
      So the area that is not carpeted is $336 \text{ ft}^2 - 153 \text{ ft}^2 = \underline{183 \text{ ft}^2}$.

   c. At first, the store would earn $10 \times \$3.50 = \$35$ a day.
      After they raise the price, they earn $7 \times \$4.50 = \$31.50$ a day.
      The store makes more money selling the bars at $\underline{\$3.50}$.

4. a. Guess and check. Think of numbers between 30 and 60 in the table of nine, and then see if that number plus 2
      is in the table of 7. $9 \times 6 = 54$ and 7×8 is 56. I am $\underline{55}$.

   b. With the number being between 37 and 100, and one digit being 3 times greater than the other digit,
      the only possibilities are 39, 62, and 93. 62 is not divisible by 3. I am $\underline{39}$ or $\underline{93}$.

   c. The two numbers less than 100 that are in both tables are 35 and 70. 3×5 is greater than $3 + 5$,
      but 7×0 is less than $7 + 0$. I am $\underline{70}$.

Multiplication and Estimation, pp. 23-24

1. a. $10 \times \$2.60 = \$26$, which is too high, so try $9 \times \$2.60 = \$23.40$. She can buy $\underline{9}$ energy bars.
   b. Round the price to $20. $5 \times \$20 = \$100$. Check with the exact price: $\$18.69 \times 5 = \$93.45$,
      so five is the most sets she can buy. She will get $5 \times 20 = \underline{100}$ tennis balls.
   c. Estimate: $5 \times \$5 = \$25$. Exact: $5 \times \$4.76 = \$23.80$. You can buy $\underline{5}$ tubs of yogurt.
   d. If we round to $40, $5 \times \$40 = \$200$, but since they cost more than $40, let's try $4 \times \$42.50 = \$170$.
      You can buy $\underline{4}$ chairs.
   e. Estimate: $8 \times \$6 = \$48$. Exact: $8 \times \$5.65 = \$45.20$. There is not enough to buy one more.
      She can get $\underline{8}$ kg of flour.

Multiplication and Estimation, cont.

Page 24

2. a. The speed doesn't matter for our calculation. We can estimate that 30 strides makes 30 × 5 m = 150 m, and one more stride will make 155 meters. It took _31_ strides.

 b. Remove the three zeros from both numbers to get 115 and 5. The question is then, how many times 5 makes 115? Since _20_ × 5 = 100 and _3_ × 5 = 15, 23 times 5 makes 115. This means 23 elephants weigh exactly the same as the whale, so _24_ elephants would weigh more.

 c. We could round 8 tentacles to 10 and estimate that it has 2,800 suction cups. Then subtract from that 2 × 280 = 560. The octopus has 2,800 − 560 = _2,240_ suction cups.

 d. Estimate: The science museum tickets would cost about 5 × $20 = $100. The aquarium tickets would cost about 2 × $25 + 3 × $15 = $95, so the science museum would be costlier to visit. (Since we rounded down the cost of the science museum tickets and rounded up the cost of the aquarium tickets, we know this won't change with exact calculations.)
 The exact cost of the aquarium tickets is 2 × $24.95 + 3 × $14.95 = $49.90 + $44.85 = _$94.75_.

Mixed Word Problems 1, pp. 25-26

Page 25

1. a. First estimate the cost per pillow of the set of two: $46 ÷ 2 = _$23_, so they are more expensive. She needs to buy 6 ÷ 2 = 3 sets in order to get six pillows. Her total is 3 × $45.65 = _$ 136.95_. Addition could also be used: $45.65 + $45.65 + $45.65 = _$ 136.95_.

 b. 9 × 5 = 45. She can feed her chickens for five days without opening up a new bag, but then there is only 50 − 45 = 5 pounds of food left in that bag, which is not enough to feed the chickens another day. The second feeding out of the bag will be in one day, the third feeding will be in two days, and so on. She will need to open a new bag in _5_ days.

2. Student's word problems will vary.
 a. EXAMPLE: You are going to ride your bike on a 332-mile trip. If you travel 83 miles per day, how many days will it take you?
 Answer: You can first use estimation. 80 × 4 = 320. Then check: 83 × _4_ = 332.
 $y = $ _4_   The trip will take you _4_ days.
 b. EXAMPLE: Mom bought 20 pieces of chicken. The family of five each ate three pieces. How many pieces are left?
 Answer: 20 − (5 × 3) = _5_.
 $y = $ _5_   There are _5_ pieces of chicken left.

Page 26

3. a. It will cost 4 × $5.50 + $44 + $120 = _$ 186_ to trim and shoe all four feet, including the shoes.
 b. $44 ÷ 4 = $11. $120 ÷ 4 = $30. It will cost $11 + $30 + $5.50 = _$ 46.50_ to trim and shoe one hoof.

4. They drove an extra 140 − 62 − 62 = 16 miles. Since that 16 miles was from driving four times around the loop, the loop was 16 ÷ 4 = _4_ miles.

Puzzle Corner: Since the first egg is laid 25 hours into the month, it is already the second day. Then, since the eggs are laid a little more than a day apart, it cannot lay 30 eggs in the remaining 29 days.
To determine exactly how many eggs the chicken did lay, first multiply 30 days by 24 hours in a day: 30 × 24 = 720 hours. Then see how many times 25 will go into 720. 25 goes into 100 four times, and there are seven hundreds; 4 × 7 = _28_, so 25 goes into 700 exactly 28 times. Now, 25 won't go into the remaining 20 hours, so in 30 days the chicken can lay _28_ eggs.
If the chicken had laid the first egg five hours earlier, there would have been just enough time for the chicken to lay another egg within the 30 days.

Dealing with Time, pp. 27-28

Page 27

1.

a. Add the minutes first. 17 + 20 = 37. This is less than 60, so add the hours. 4 + 3 = 7. The airplane will land at _7:37 PM_.	b. 10:46 − 1:30 = _9:16_. She started at _9:16 AM_.
c. From 12:25 to 1:10 is 35 + 10 = 45 minutes. Eighty minutes is 1 hour and 20 minutes, or 1:20. So their trip was 2:00 + 2:00 + 0:45 + 1:20 + 2:00 = _8:05 or 8 hours and 5 minutes_.	d. She spends 30 × 7 = _210 minutes or 3 hours and 30 minutes_ cleaning per week.
e. She will have saved $3 × 365 = _$1,095_.	f. He should leave at 5:30 − 0:23 = _5:07_.
g. Wednesday, Thursday, Friday, and Saturday is four days. It took him 4 × 2:30 + 0:15 = _10:15 or 10 hours and 15 minutes_ to finish the painting.	

Page 28

2.

Larry's pasture rotation	
July 17 - July 21	Pasture 1
July 22 - July 26	Pasture 2
July 27 - July 31	Pasture 3
August 1 - August 5	Pasture 4
August 6 - August 10	Pasture 5
August 11 - August 15	Pasture 6
August 16 - August 20	Pasture 7
August 21 - August 25	Pasture 8

a. They go through seven other pastures for five days each, so each pasture gets 7 × 5 = _35_ days rest.

b. Seven weeks is 49 days. Each pasture adds five days to the rotation schedule. With 10 pastures, each pasture would get 9 × 5 = 45 days of rest, which isn't quite enough. With 11 pastures, it will work: each pasture will get 10 × 5 = 50 days of rest.

c. 16 pastures and three days per pasture means that after the sheep leave any pasture, they will have 15 other pastures to go through before coming back. Therefore, each pasture will get 15 × 3 = 45 days of rest.

Practice 24-Hour Time, Version 1, p. 29

Page 29

1. From 16:21 to 17:00 is 60 − 21 = 39 minutes. So from Fresno to San Francisco is 39 + 26 = _65 minutes or 1 hour and 5 minutes_.
From 20:50 to 24:00 is 3:10 hours. So from San Francisco to Copenhagen is 3:10 + 7:40 = _10:50 or 10 hours and 50 minutes_.

2. a. In 24-hour time it is _23:12_. He is in _Frankfurt_.
b. It will be 0:48 + 4:10 = _4:58 or 4 hours and 58 minutes_ until he reaches Stockholm.

Practice 24-Hour Time, Version 1, cont.

Page 29

3. She should take _Route 1_, as the flight from San Francisco to Copenhagen is from 20:50 to 07:40.

4. Route 1, from 16:21 to 24:00 is 7:39. 7:39 + 11:55 = 19:34 or 19 hours and 34 minutes.
   Route 2, from 09:47 to 24:00 is 14:13. 14:13 + 04:10 = 18:23 or 18 hours and 23 minutes.
   He should take _Route 2_.

Practice 24-Hour Time, Version 2, p. 30

Page 30

1. From Fresno to San Francisco is 39 + 26 = _65 minutes or 1 hour and 5 minutes_.
   Copenhagen is 9 hours ahead of San Francisco, so add 9 hours to the time in San Francisco, or
   subtract 9 hours from the time in Copenhagen. 24:00 − 20:50 = 3:10. 16:40 − 9:00 = 7:40.
   From San Francisco to Copenhagen is 3:10 + 7:40 = _10:50 or 10 hours and 50 minutes_.

2. a. In 24-hour time, it is _13:10_.
      He has just arrived in _Denver_.
   b. Stockholm is 8 hours ahead of Denver, so add 8 hours to the time in Denver or subtract 8 hours
      from the time in Stockholm. 13:10 + 8:00 = 21:10. 24:00 − 21:10 = 2:50.
      It will be 2:50 + 13:10 = _16:00 or 16 hours_ until he reaches Stockholm.

3. She should take _Route 1_. It leaves San Francisco at 20:50 and arrives in Copenhagen at the
   equivalent of 07:40 in San Francisco.

4. For both routes, there is a 9-hour time difference between Fresno and Stockholm. Add 9 hours
   to the time in San Francisco, or subtract 9 hours from the time in Stockholm.
   Route 1, from 16:21 to 24:00 is 7:39. 7:39 + 11:55 = 19:34 or 19 hours and 34 minutes.
   Route 2, from 18:47 to 24:00 is 5:13. 5:13 + 13:10 = 18:23 or 18 hours and 23 minutes.
   He should take _Route 2_.

Units of Length, p. 31

Page 31

1. a. First, convert feet to inches. The crates are 12" + 7" = 19" tall. The ceiling is 8 × 12" = 96" tall.
      Estimate that 5 × 20" = 100". The actual height of five crates is 5 × 19" = 95". You can stack
      five crates.

   b. She makes the trip 2 × 5 = 10 times in a week. So she walks 10 × 1,410 = _14,100_ feet in a school week.
      Round the length of a mile to 5,000 feet. 3 × 5,000 = 15,000, so it is *about* _3 miles_.

   c. The longest caterpillar is 6 cm = 60 mm long. It is 60 mm − 18 mm = _42 mm_ longer than the shortest caterpillar.
      The "train of caterpillars" would be 60 + 40 + 37 + 24 + 18 = _179 mm_ long.

   d. Charlie ran 4 × 3 = 12 feet before he was caught. Then he was dragged another 3 × 3 + 2 = 11 feet.
      He traveled 12 ft + 11 ft = _23_ feet with muddy feet.

Mixed Word Problems 2, pp. 32-33

Page 32

1. a. Ranch B had *about* 2,900 − 1,950 = _950_ more cattle. (Answers may vary slightly.)
   b. The number of cattle at Ranch A was at its lowest in year 2, at *about* 1,250. It was double that amount
      in _year 6_ at *about* 2,500.
   c. The difference was the greatest in _year 1_.

Mixed Word Problems 2, cont.

Page 32

2. The table below shows the heights of both drones over time.

Time	start	1 min	2 min	3 min	4 min
Tim's drone	1,525	1,490	1,425	1,428	1,407
Bill's drone	1,342	1,310	1,364	1,428	1,474

They would have been at the same height 3 minutes after the start of the flight.

Page 33

3. a. $16 \times 12 = 192$ cents, or \$1.92. $14 \times 14 = 196$ cents, or \$1.96. The first watermelon would be a little cheaper.

   b. Five kilometers is 5,000 meters. Each segment of her run+walk is $800 + 100 = 900$ meters. Think how many times 900 will go into 5,000. You can remove two zeros off the end of each number to make it easier (to get 50 and 9). Now, $6 \times 9 = 54$, which is too high, so try $5 \times 9 = 45$, or $5 \times 900 = 4,500$. That works. Of each of those five segments, she would have run 800 meters and then she would have also run the last $5,000 - 4,500 = 500$ meters. So in total, she actually ran $5 \times 800 + 500 = $ _4,500 meters_ . Or, you can calculate that she walked $5 \times 100 = 500$ meters, and therefore she ran $5,000 - 500 = $ _4,500 meters_ .

   c. He still needs to save $\$1,559 - \$342 = \$1,217$. Each week he can save $5 \times \$140 - \$485 = \$215$. Round \$1,217 and \$215 to the nearest hundred and estimate: _6_ $\times \$200 = \$1,200$. Then check the exact numbers: $6 \times \$215 = \$1,290$. He could not have saved enough in only five weeks. It will be _6 weeks_ before he can buy the camera.

Division Warm-Up, pp. 34-35

Page 34

1. a. $36 \div 4 = $ _9._ There are nine deer.
   b. $4 \times 4 = $ _16_ There are sixteen pieces of apple.
   c. $60 \div 12 = $ _5_ There are five bags of apples.

   d. $\$20 \div \$4 = $ _5_ He bought five pineapples.
   e. $246 - 13 = $ _233_ There are 233 books left on the bookcase.
   f. $4 \times \$5.50 = $ _\$22.00_ You spent \$22.00 on the four books.

   g. $\$24 \div 8 = $ _\$3_ Each notebook costs \$3.
   h. $200 \div 4 = $ _50_ There are fifty pencils in each box.
   i. $12 \div 3 = $ _4_ They each get four squares of chocolate.

Page 35 35

2. The cost of the pizza is $3 \times \$14$. It is divided among six friends, so the correct calculation is _$3 \times 14 \div 6 = x$_ . $3 \times 14 = 42$; $42 \div 6 = $ _7_ . Each friend paid _\$7_ .

3. a. _8_ $\times 12 = 96$; $100 \div 12 = 8$ with a remainder of 4. He gets _8_ full bags of apples. There are _4_ apples left.
   b. From 9:08 AM to 11:21 AM is 2 hours and 13 minutes, or 133 minutes. $133 \div 4 = 33$ with a remainder of 1. She answered about _33_ emails.
   c. The discount on the granola is $1 - 2/3 = 3/3 - 2/3 = 1/3$. To find 1/3 of \$24, divide: $24 \div 3 = $ _8_ . The granola is discounted by _\$8_ .
   d. $50 \div 4 = 12$ with a remainder of 2. The bag of food will last _12_ days.
   e. For seven plants in a row: $29 \div 7 = 4$ with a remainder of 1. He will get _4_ complete rows, with _1_ plant in the incomplete row.
   For six plants in a row: $29 \div 6 = 4$ with a remainder of 5. He will get _4_ complete rows, with _5_ plants in the incomplete row.

An Expression for a Word Problem, pp. 36-37

Page 36

1. a. The expression can be written $9 \times 12 - 28 - 37 - 4$ or $9 \times 12 - (28 + 37 + 4)$.
   $108 - 69 = \underline{39}$. He has $\underline{39}$ golf balls left.

   b. $28 \div 4 + 36 \div 6 = 7 + 6 = 13$. One package of each costs $\underline{\$13}$.

Page 37

2. a. v. $4 + 36 - 7 = \underline{33}$. He has $\underline{\$33}$ now.
   b. iii. $36 \div 4 + 7 = \underline{16}$. One puppy now weighs $\underline{16}$ pounds.
   c. ii. $(4 + 7) \times 7 + 36 = \underline{113}$. He spent $\underline{\$113}$ in total.
   d. iv. $36 - 7 \times 4 + 36 = \underline{44}$. He has $\underline{\$44}$ now.
   e. vi. $36 + 7 \times 4 = \underline{64}$. She has laid $\underline{64}$ eggs now.
   f. i. $(36 - 4 \times 7) \div 4 = \underline{2}$. The other side was $\underline{2}$ feet long.

Division Practice, pp. 38-40

Page 38

1. See the long division on the right.
   The hummingbirds are drinking $\underline{250}$ grams in a day.

$$
\begin{array}{r}
250 \\
3 \overline{)750} \\
-6 \\
\hline
15 \\
-15 \\
\hline
0
\end{array}
$$

2. $(89 - 11) \div 6 = 78 \div 6 = \underline{13}$.
   One towel cost $\$ \underline{13}$.

3. See the long division on the right.
   He has $\underline{2,683}$ cows.

4. $256 - 8 \times 4 = 256 - 32 = \underline{224}$
   or $256 \div 8 = 32$; $32 - 4 = 28$; $28 \times 8 = \underline{224}$.
   She got $\underline{224}$ eggs this year.

$$
\begin{array}{r}
2683 \\
3 \overline{)8049} \\
-6 \\
\hline
20 \\
-18 \\
\hline
24 \\
-24 \\
\hline
09 \\
-9 \\
\hline
0
\end{array}
$$

Page 39

5. $141 - 48 \times 2 = 141 - 96 = 45$; $45 \div 3 = \underline{15}$.
   $\underline{Fifteen}$ practices lasted three hours.

6. First we will subtract the cost of the two jars of
   sauerkraut from the total:
   $\$94.55 - 2 \times \$18.90 = \$94.55 - \$37.80 = \$56.75$.
   Then we will divide that by five to get the cost
   of one jar of jam. See the long division on the right.
   Each jar of jam cost $\$ \underline{11.35}$.

$$
\begin{array}{r}
11.35 \\
5 \overline{)56.75} \\
-5 \\
\hline
17 \\
-15 \\
\hline
25 \\
-25 \\
\hline
0
\end{array}
$$

Page 39

7. Each truck starts with $7,830 \div 6 = 1,305$ kg of blueberries.
   (See the division on the right.)

   A third of a truckload that got dumped to Mathy's
   friend's house was $1,305$ kg $\div 3 = 435$ kg.

   $7,830 - 435 = \underline{7,395}$.
   Mathy got $\underline{7,395}$ kg of blueberries.

$$
\begin{array}{r}
1305 \\
6 \overline{)7830} \\
-6 \\
\hline
18 \\
-18 \\
\hline
030 \\
-30 \\
\hline
0
\end{array}
$$

Division Practice, cont.

8. Answers will vary. For example:
  a. Sherry split 123 strawberries between six of her friends, with each getting an equal amount of whole berries.
     How many strawberries did each one get, and how many were left over?
     Answer: 123 ÷ 6 = 20 R3. Each friend got _20_ strawberries, and there were _3_ left over.

  b. Mrs. Teacher divided 48 crayons equally between 12 students, but gave Stuart three extra crayons.
     How many crayons did Stuart get?
     Answer: 48 ÷ 12 + 3 = _7_. Stuart got _7_ crayons.

Puzzle Corner: She uses 95 × 50 lb = 4,750 lbs of chicken food in a year.
We can divide that by 365 to find out how many pounds she uses in a day.

```
        0 0 1 3
365)4 7 5 0
   - 3 6 5
     1 1 0 0
    - 1 0 9 5
            5
```

The division doesn't come out exactly even, but since it is very close, we can say she uses
13 lb of chicken feed per day. Since each chicken eats 1/2 lb, this means she has _26_ chickens.

Estimation and multiplication could also be used. One chicken eats 365 × 1/2 lb = 182.5 lbs
in a year. And in total, she uses 4,750 lb in a year. Rounding 182.5 to 200 and 4,750 to 5,000,
we can estimate that she has 5,000 ÷ 200 = 50 ÷ 2 = 25 chickens. Now, check with exact numbers:
multiply 25 × 182.5 lb = 4,562.5 lb. Add another 182.5 lb to that: 4,562.5 lb + 182.5 lb = 4,745 lb, so again,
she has _26_ chickens.

Units of Volume, p. 41

1. a. They are drinking 2 × 300 ml + 3 × 250 ml = 1,350 ml of juice a day. Four liters is 4,000 ml.
     1,350 + 1,350 = 2,700; 2,700 + 1,350 = 4,050. So the jug will last *about* _three_ days.

  b. Fifteen gallons is 15 × 4 = 60 quarts. 60 ÷ 5 = 12; 12 × 4 = _48_. 4/5 of the tub full is _48_ quarts.

  c. She would eat a full pint every two days. There are two pints in a quart, so she would eat a quart of yogurt in
     2 × 2 = _4_ days. There are four quarts in a gallon, so a gallon of yogurt would last her 4 × 4 = _16_ days.
     Two quarts of yogurt will last her 8 days, so it will cost 2 × $5.35 = _$10.70_ for enough
     yogurt to last at least a week.

  d. 3 oz + 1 oz = 4 oz = 1/2 cup. 8 × 1/2 cup = 4 cups. It will take 8 × 3 = _24_ oz of water
     and 8 × 1 = _8_ oz of concentrate to make 4 cups of juice.

Finding the Average, pp. 42-43

1. a. 378 ÷ 9 = _42_.   b. 49 × 6 = _294_.
     His average speed was _42_ miles per hour.   She drove _294_ miles.

  c. Plant A: 22 + 23 + 18 = 63; 64 ÷ 3 = 21. The average watermelon weighs 21 pounds.
     Plant B: 23 + 20 + 16 + 21 = 80; 80 ÷ 4 = 20. The average watermelon weighs 20 pounds.
     Plant A made bigger watermelons on average. They were about 21 − 20 = _1_ pound larger.

2. 4,032 + 9 × 7 = 4,032 + 63 = _4,095_.
   The trees produced _4,095_ mangoes this year.

3. 128 + 98 + 105 + 114 + 120 = 565; 565 ÷ 5 = 113 inches. Since 9 × 12 + 5 = 113, this is 9 feet 5 inches.
   Their average wingspan is _9 feet 5 inches_.
   Add 137 to the original total: 565 + 137 = 702; 702 ÷ 6 = 117 inches. Since 9 × 12 + 9 = 117, this is 9 feet 9 inches.
   The average wingspan is now _9 feet 9 inches_.

Puzzle Corner: Each teabag costs $5.40 ÷ 18 = $0.30. Using one teabag per day, she will spend 30 × $0.30 = _$9.00_
on tea in a month.

Units of Weight, p. 44

Page 44

1. a. 120×50 g $= 6,000$ g.
   6,000 grams is _6_ kilograms of blueberries.

   b. $482 + 493 + 482 + 471 + 488 + 486 = 2,904$; $2,904 \div 6 = 484$. The average weight is _484 g_.
   $6,045 - 493 = 5,552$ g. It has gained 5,552 g, or _5 kg 552 g_, since birth.

   c. 9 pounds 4 ounces is 9 1/4 pounds. $50 - 9\ 1/4 = 40\ 3/4$, so the lamb needs to gain a little over 40 pounds.
   $4 \times 2\ 1/2 = 10$, so it gains 10 pounds every four weeks. It needs to gain four times that amount. $4 \times 4 = 16$,
   so it will take _a little over 16 weeks_ for the lamb to reach 50 pounds.

   d. Two pounds is $2 \times 16 = 32$ ounces. $3 \times 12 = 36$, so he needs _3_ packages of cream cheese.
   There will be $36 - 32 =$ _4 ounces_ of cream cheese left over.

Divisibility and Factors, pp. 45-46

Page 45

1. a. $7 \times 5 = 35$, which is too low, so the number must have another factor. Start with the lowest: $2 \times 35 =$ _70_.
   b. You can check each number between 90 and 100 until you find the one that's divisible by all three factors.
   Or multiply $2 \times 3 \times 4 = 24$. Then multiply by an additional factor until you reach the target number. In this
   case, $4 \times 24 =$ _96_.

   c. The prime needs to be an odd number. 91 is divisible by 7. 93 and 99 are divisible by 3. 95 is divisible by 5.
   The only prime number between 90 and 100 is _97_.

   d. The number is _36_. Its factors are: 1, 2, 3, 4, 6, 9, 12, 18, 36.

   e. The number is _84_. Its factors are: 1, 2, 3, 4, 6, 7, 12, 14, 21, 28, 42, 84.

   f. The number is _49_, with factors 1, 7, and 49. For any number above 4 to have exactly three factors, it must be an odd
   number. Two of the factors will be 1 and the number itself, and the third factor is some number that when multiplied
   by itself gives the number in question. In other words, the number has to be an odd square number. Since $7 \times 7 = 49$,
   49 works, and is the only such number between 40 and 70.

Page 46

2. $3 \times 17 = 51$. The other two numbers are _1_ and _51_.

3. Luis is not correct. 31 and 37 are the only primes between 30 and 40, and are more than two numbers apart.
   Three, five, and seven are the only three consecutive odd numbers that are primes. Any higher three consecutive odd
   numbers will have at least one of them divisible by three.

Page 46

4. Numbers that are divisible by both 4 and 6 are not simply multiples of $4 \times 6 = 24$. Instead, all multiples of 12 are
   divisible by both 4 and 6. The first such multiple above 60 is 72, which is divisible by 8, so it won't work. The
   second one is 84. It is not divisible by 8. The last such multiple is 96, which again *is* divisible by 8.
   Therefore, the number is _84_.

5. _21_   Factors: 1, 3, 7, 21
   _22_   Factors: 1, 2, 11, 22
   _26_   Factors: 1, 2, 13, 26
   _27_   Factors: 1, 3, 9, 27

6. Jessica is correct. For a number to have an odd number of factors, it must be a perfect square –the product
   of a number multiplied by itself, for example 5×5. The perfect squares less than 50 are 0, 1, 4, 9, 16, 25, 36, and 49.
   Of those, 0 and 1 only have one factor. Then, 4, 9, 25, and 49 have only three factors. Sixteen has five factors, but is
   not divisible by 3. Lastly, 36 has more than five factors.

Puzzle Corner: _25_ (1, 5, 25) and _49_ (1, 7, 49) are the only two 2-digit numbers that have exactly three factors.

<u>Page 47</u>

1. a. The area of the whole house is 40 ft × 38 ft = _1,520 ft^2_.

   b. 26 + (40 − 20) + 38 + (40 − 20 − 15) + 12 + 15 = 26 + 20 + 38 + 5 + 12 + 15 = _116 ft_.
   It may be noticed that the inner walls of Room B are both parallel with the outer walls, making them the same lengths, so the perimeter could be calculated as 2 × (40 − 20) + 2 × 38 = _116 ft_.

   c. First calculate the area of the full rectangle, including Room B: 38 ft × (40 − 20) ft = 38 ft × 20 ft = 760 ft^2.
   Then subtract the area of Room B: 760 ft^2 − 180 ft^2 = _580 ft^2_. The area of Room A is _580 ft^2_.

2. The tiles are 1 ft × 1 ft, or 1 square foot each, so the number of tiles needed will be the same as the area of the room in square feet. 15 ft × 17 ft = 255 ft^2. He will need _255_ tiles for the room.

3. First determine the length of the other side(s) of the porch: 21 m^2 ÷ 7 m = 3 m. She will need 3 m + 3 m + 7 m = _13 m_ of screen.

<u>Page 48</u>

4. Answers will vary. Check the student's work. For example:
   She could have an 18-in wide bed across one 12-ft end of the garden, and then 18-in wide paths and 3-ft wide beds with an 18-in wide path across the ends of them. The area would be:
   1.5 ft × 12 ft + 6 × (3 ft × 10.5 ft) = 18 ft^2 + 6 × 31.5 ft^2 = 18 ft^2 + 189 ft^2 = _207 ft^2_.

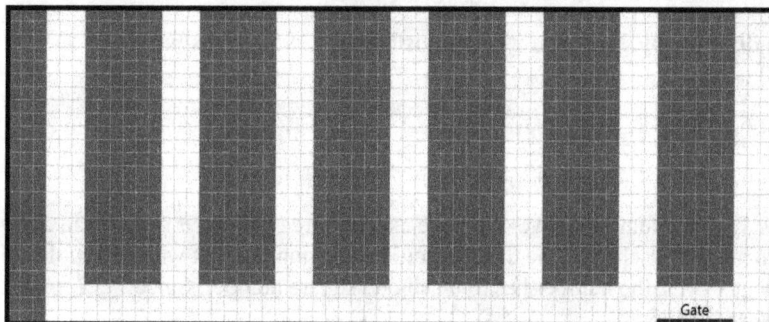

Another option would be to have an 18 in-wide bed on one side of the 30-ft length, then 3 ft-wide beds running the same direction, with 18-in wide rows in between and along the other 30-ft side. Then have a 2-ft wide aisle splitting the 3-ft wide rows in half lengthwise. The area of the beds would be:
1.5 ft × 30 ft + 4 × (3 ft × 14 ft) = 45 ft^2 + 4 × 42 ft^2 = 45 ft^2 + 168 ft^2 = _213 ft^2_.

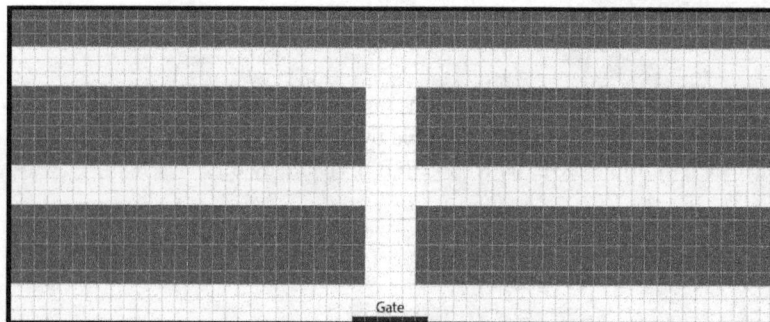

Angles and Shapes, pp. 49-50

Page 49

1. a. Angle ACD is 90°, so the sum of the other two angles in the triangle must be 90°, so it is not possible for one of them to be 117°.
   Hailey probably looked at the numbers starting from the wrong side of the protractor.

   b. The first angle is 90°. The second angle is 1/3 of that, or 90 ÷ 3 = 30°. The third angle is half of the second, or 30 ÷ 2 =15°. The angle sum is then 90 + 30 + 15 = _135°_.

   c. We know that in the multiplication table 8 × 8 = 64, so the first angle has to be 64°.
   Because the other two angles are the same, they must be (360 − 64) ÷ 2 = 296 ÷ 2 = 148°.
   The three angles measure 64°, 148°, and 148°.

Page 50

2. He is not correct. It is an obtuse triangle, because it has an angle greater than 90° (the top angle). *All* angles in an acute triangle must be less than 90°.

3. a. Student drawings will vary. Each triangle that is formed must have an angle greater than 90°. See an example on the right:

   b. The quadrilateral must be a rectangle or a square, so that the two triangles formed will each have an angle of 90°. Example:

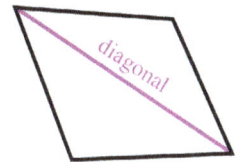

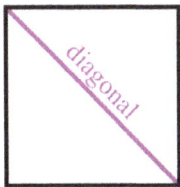

4. It is a right triangle:

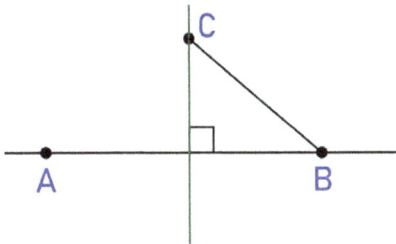

Fraction Fun, pp. 51-52

Page 51

1. a. Mom can make oatmeal _5_ times, using one box of oats, and there will be 1 1/4 cups of oats left.

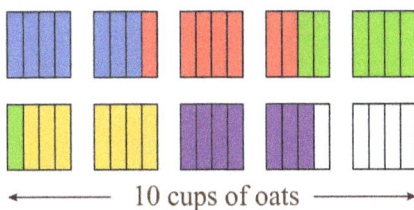

← 10 cups of oats →

   b. The two 1/2 cups make one whole cup, so she used 5 + 2 + 1 = _8_ cups of flour in total.
   c. 6 − 2 = 4 and 3/4 − 1/4 = 2/4 = 1/2. Julie ran _4 1/2_ fewer laps.

Fraction Fun, cont.

Page 52

2.

a. 1 1/3 cups is 4/3 cups and $3 \times \frac{4}{3} = \frac{12}{3} = 4$. You will need _4_ cups of butter.	b. 1/4 of an hour is 1/4 of 60 minutes = 15 minutes. 1 hour and 15 minutes after 4:10 is 5:25. He finished at _5:25_.
c. $5 \times \frac{1}{2} = \frac{5}{2} = 2\frac{1}{2}$ The tomato plant is _2 1/2 meters_ tall now.	d. 2 1/5 + 1 3/5 = 3 4/5. 5 − 3 4/5 = _1 1/5_ The third cat got _1 1/5_ fish.
e. 3/4 + 2/4 = 5/4 = 1 1/4 2 × 1 1/4 = 2 2/4 = _2 1/2_ The deer traveled _2 1/2_ miles.	

Puzzle Corner: In 1/4 of an hour she will walk 1/4 of 4 miles, or _1 mile_.
Since she takes 15 minutes to walk a mile, she will take 1/3 that much time, or _5 minutes_, to walk 1/3 of a mile.

Fractions and Division, pp. 53-54

Page 53

1. a.
```
       1 9
   9 ) 1 7 1
      - 9
       8 1
      - 8 1
         0
```
They have driven 4/9 of the way, so have 9/9 − 4/9 = 5/9 left to go. From long division, we have determined that 1/9 of 171 is 19. So they have 5 × 19 = _95_ km to go to get to the beach.

b. Answers may vary. Please check the student's answer.
   If the distance is rounded to 180 km, they are about 60/180 = 6/18 = _1/3_ of the way home.

c. 3 of 12, or 3/12 = _1/4_ of the birds are robins.

   1/3 of 12 is 4. There were _4_ finches.
   12 − 3 − 4 = _5_ of the birds weren't robins or finches.

d. Three out of ten slices is 3/10 of the cheesecake. Each slice weighs 850 ÷ 10 = 85 g. Three slices would weigh 85 × 3 = _255_ g.

Page 54

2. a. To find 1/3 of a number, divide by 3.
```
          4. 2 0
   3 ) 1 2.6 0
      - 1 2
         0 6 0
         - 6 0
             0
```
   Her total cost was
   12.60 + 4.20 = $ _16.80_.

b. One-fourth of 16 is 16 ÷ 4 = _4_. The first piece he sawed off is _4 feet_ long.
   The rest of the plank was 16 − 4 = 12 feet long. 12 ÷ 3 = _4_. The other three pieces are also _4 feet_ long.

c. Her sales for the first day were 52 × $2 = $104. To find 1/5, divide by five: $104 ÷ 5 = $20.80.
   She has $104 − $20.80 = $ _83.20_ left.

d. First find the total amount of time in minutes: 2 × 60 + 20 = 140. 1/7 of 140 is 140 ÷ 7 = 20.
   She spends 20 × 5 = 100 minutes working, and 140 − 100 = 40 minutes playing. She spends
   100 − 40 = _60_ minutes more working than playing.

Decimals, pp. 55-56

Page 55

1. a. $\underline{1.5}$ feet is 1 foot 6 inches, so is shorter.   b. The perimeter is 18.2 m + 18.2 m + 21.5 m + 21.5 m = $\underline{79.4 \text{ m}}$.

   c. $5.29 + $5.29 + $5.29 = $15.87. His change is $20 − $15.87 = $\underline{\$4.13}$.

   d. 2.4 kilometers is 2,400 meters. Philippa walked 2,400 m − 1,400 m = $\underline{1,000}$ meters more.

   e. The perimeter of the rectangle is 5.74 m + 5.74 m + 2.8 m + 2.8 m = 17.08 m. The perimeter of the square is 4.3 m + 4.3 m + 4.3 m + 4.3 m = 17.2 m. The perimeter of the $\underline{\text{square}}$ is 17.2 m − 17.08 m = $\underline{0.12 \text{ m}}$ larger.

   f. Round 18.6 m to 20 m, and estimate that it will take 5 trips total. But since since each trip is less than 20 m, it will take one more trip to go at least 100 m. 18.6 m + 18.6 m + 18.6 m + 18.6 m + 18.6 m + 18.6 m = 111.6 m. It will take 6 − 2 = $\underline{4}$ more trips (or two more round trips) for the squirrel to go over 100 m.

Page 56

2. a. The total weight is 0.85 lb + 0.62 lb + 5.2 lb + 3.9 lb = $\underline{10.57 \text{ lb}}$.

   b. 1/2 is 0.5. Daniel has to go 0.6 mi − 0.5 mi = $\underline{0.1 \text{ mi}}$ farther.
   Daniel bicycles 0.6 mi + 0.6 mi + 0.6 mi + 0.6 mi = $\underline{2.4 \text{ mi}}$.
   Harold bicycles 0.5 mi + 0.5 mi + 0.5 mi + 0.5 mi = $\underline{2.0 \text{ mi}}$.

   c. 1.82 kg is 1,820 g. 400 g + 400 g + 400 g + 400g = 1,600 g. She can make $\underline{4}$ loaves of bread. There is not enough flour left for another loaf.

   d. Their combined weight initially is 8.6 lb + 7.8 lb = 16.4 lb. You can add 5.2 lb to that total six times, or add 5.2 lb + 5.2 lb = 10.4 lb, and then add 10.4 lb three times. The lambs' total combined weight after three weeks is 16.4 lb + 10.4 lb + 10.4 lb + 10.4 lb = $\underline{47.6 \text{ lb}}$.

Puzzle Corner: Dad, Kirsten, and Monica ate a total of 1.5 − 1 = 0.5, or 1/2 gallon of ice cream. There are 2 cups per pint, 2 pints per quart, and 2 quarts in half a gallon, so they ate 2 c × 2 × 2 = 8 cups. Kirsten and Monica each ate 1.5 cups at 2 meals, so 1.5 c + 1.5 c + 1.5 c + 1.5 c = 6 c. Dad ate 8 c − 6 c = $\underline{2 \text{ cups}}$ of ice cream.

Mixed Word Problems 3, pp. 57-58

Page 57

1. a. She will have walked Monday, Wednesday, Friday, and Sunday, or 4 days. The total distance she will have walked is 4 × 1,530 m = $\underline{6,120 \text{ m}}$.

   b. $23.60 ÷ 4 = $5.90 and $47.20 ÷ 8 = $5.90, so each shirt costs $5.90. A pack of six shirts would cost $5.90 × 6 = $\underline{\$35.40}$. A pack of five shirts would cost $5.90 × 5 = $\underline{\$29.50}$.

   c. Five-fold is the same as five times, so now he has 5 × 14 = 70 rabbits. Next year, he expects to have 3 × 70 = $\underline{210}$ rabbits.

   d. $\dfrac{5}{5} - \dfrac{1}{5} = \dfrac{4}{5}$
   He now has 4/5 of his savings left. So 1/5 is $20 ÷ 4 = $\underline{\$5}$. The magnifying glass cost $\underline{\$5}$.

Page 58

2. Yes, that is enough information to figure out what the area is.

One side of the pasture is 29 m × 3 = 87 m. The area of the pasture is 87 m × 87 m = 7,569 m^2.

Here is a sketch for one side of the square. The gate and the 26-m stretch are not to scale.

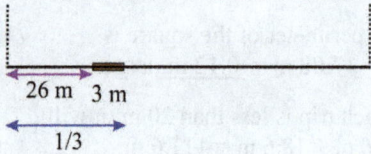

26 m   3 m
1/3

3. One side of the new pasture will match the length of the side of the original pasture, or 87 m. So the sum of the other two sides will be 251 m − 87 m = 164 m. This means one of the sides will be 164 ÷ 2 = 82 m. The new pasture will be 87 m × 82 m .

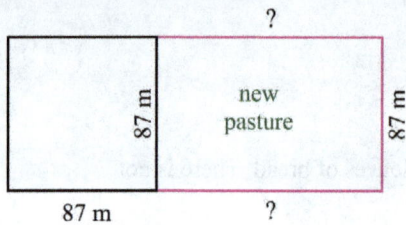

More from math MAMMOTH

Math Mammoth has a variety of resources to fit your needs. All are available as economical downloads, and most also as printed copies.

- **Math Mammoth Light Blue Series**
  A complete curriculum for grades 1-8. Each grade level includes two student worktexts (A and B), which contain all the instruction and exercises all in the same book, answer keys, tests, cumulative reviews, and a worksheet maker. International (all metric), Canadian, and South African versions are also available.
  https://www.MathMammoth.com/complete-curriculum
  https://www.MathMammoth.com/international/international
  https://www.MathMammoth.com/canada/
  https://www.MathMammoth.com/south_africa/

- **Math Mammoth Skills Review Workbooks**
  These workbooks are intended to be used alongside the Light Blue series full curriculum, and they provide additional review to the topics studied in the main curriculum, in a spiral manner.
  https://www.MathMammoth.com/skills_review_workbooks/

- **Math Mammoth Blue Series**
  Blue Series books are topical worktexts for grades 1-8, containing both instruction and exercises. They cover all elementary math topics from 1st through 8th grade and some for 8th grade. These books are not tied to grade levels, and are thus great for filling in gaps.
  https://www.MathMammoth.com/blue-series

- **Make It Real Learning**
  These activity workbooks concentrate on answering the question, "Where is math used in real life?" The series includes various workbooks for grades 3-12.
  https://www.MathMammoth.com/worksheets/mirl/

- **Review Workbooks**
  Workbooks for grades 1-8 that provide a comprehensive review of one grade level of math —for example, for review during school break or summer vacation.
  https://www.MathMammoth.com/review_workbooks/

Free gift!

- Receive over 350 free sample pages and worksheets from my books, plus other freebies:
  https://www.MathMammoth.com/free/

Lastly...

- Inspire4 is an inspirational website for the whole family I've been privileged to help with:
  https://www.inspire4.com

www.ingramcontent.com/pod-product-compliance
Lightning Source LLC
Chambersburg PA
CBHW051231200326
41519CB00025B/7335